DR RICHARD THO[illegible] professor at the Unive[illegible] [illegible], environment and conservation. Widely travelled, he has a longstanding interest in indigenous and sustainable farming. He was introduced to the work of Rudolf Steiner at an early age, although his full involvement with biodynamics dates from 1990 when he began to participate in training programmes and workshops at Emerson College, Sussex. In 1996 he began a biodynamic extension programme in Sri Lanka, for which he published a book, updated in 2007. Since 2001 he has been an inspector for the Biodynamic Association's Demeter and Organic Certification in the UK. In 2003 he produced an edited selection of Steiner's work relating to agriculture. He is currently a council member of the Biodynamic Agricultural Association, and lives in Ross-on-Wye, Herefordshire.

COSMOS, EARTH AND NUTRITION

The Biodynamic Approach to Agriculture

Richard Thornton Smith

Sophia Books

Sophia Books
Hillside House, The Square
Forest Row, RH18 5ES

www.rudolfsteinerpress.com

Published by Sophia Books 2009
An imprint of Rudolf Steiner Press

A catalogue record for this book is available from the British Library

ISBN 978 1 85584 227 4

Cover by Andrew Morgan Design
Typeset by DP Photosetting, Neath, West Glamorgan
Printed and bound in Malta by Gutenberg Press Ltd.

The paper used for this book is FSC-certified and totally chlorine-free. FSC (the Forest Stewardship Council) is an international network to promote responsible management of the world's forests.

Contents

Preface

There has been growing interest in biodynamics in recent years. For a long time the organic farming and gardening movement was dismissed as marginal and 'alternative', but escalating concerns about the environment, health, food quality, animal welfare and related issues have raised the profile of organics in a way few could have imagined 30 years ago. As a branch of organic agriculture, it is scarcely surprising that biodynamics has emerged more into public consciousness.

Biodynamics was never narrowly focused on agricultural techniques. It was conceived as a new way of thinking about farming, nutrition and the world of nature. Originating from a series of eight lectures on agriculture given by Dr Rudolf Steiner in 1924, it offers a new holistic outlook that frees agriculture and science from the limits of a purely materialist philosophy.

Like others before it, the present book supports practical biodynamic farming and gardening. Those already involved with biodynamics should find among its broad range of topics much to deepen their understanding of the subject. They will also detect the author's concern to promote innovation and for biodynamics to move with the times. Those new to biodynamics, but perhaps already committed to an organic philosophy, may have little idea of Rudolf Steiner's immense contribution to knowledge. A major task of this book is therefore to set out the fundamentals on which not only biodynamics but also the wider organic movement depend. To meet this need, the book aims wherever possible to create a bridge between mainstream science and Steiner's insights and suggestions, and to offer the wider organic and agro-ecological movement a firmer basis for acknowledging biodynamic concepts.

While Steiner's *Agriculture Course* remains the cornerstone, those working with biodynamics have found it essential to study his other lectures and books to gain better foothold with the content of the agriculture lectures, and to extend their scope. This is a long, ongoing process. Yet there is an urgency about getting to grips with the current disruptions to climatic, ecological and economic conditions, and this concern has provided the motivation for the present volume.

Considered as a group, the first four chapters lay the foundations. I first place biodynamics in the context of holistic forms of agriculture, outlining the cultural fault-line with conventional agriculture and the kinship of biodynamics with organic traditions and the wider agro-ecological movement. Chapters 2, 3 and 4 explore themes from the *Agriculture Course*—the cosmic dimension, the farm organism and the role of the soil. Crucial to the aim of this book is that we adopt a more enlightened view of how the farm or garden manages the interplay of earth and cosmos, and how soil, too often regarded as the most mundane of substances, truly sustains the whole of life.

Having widened our view of nature, the principal biodynamic practices are presented in Chapters 5 and 6. In previous literature these are mostly treated as axiomatic. Here, we engage in discussion and critique rather than straightforward description of procedures. In addition to the established biodynamic preparations, the window is opened on a range of less familiar innovations with biodynamic pedigree. Chapter 6, on the biodynamic calendar, includes much that is familiar to organic gardeners but urges reappraisal of the use of the zodiac and a clearer understanding of the opportunities a calendar may offer. Chapter 7, on seeds, represents a topic of vital importance to organic and biodynamic growers in face of challenges from the biotech world. Chapter 8 enters into the special character of water, the most vital of substances supporting life, and the way in which it is used and treated in biodynamic procedures.

The last four chapters, in their different ways, all connect with the human being. Chapter 9 examines the relationship between outer visible landscapes and our inner mental landscapes, and contends that working with nature in the biodynamic way can offer a mutually healing process for society and the earth. Chapter 10, which tackles the immense subject of food and health, will represent for many people the prime mission of biodynamic agriculture and is justification alone for probing the hidden pathways of nature explored in this book. Through discussion of modern health problems, the chapter confirms the wisdom of a holistic approach to nutrition.

Chapter 11 introduces the social element of agriculture. So far of limited impact, community involvement with agriculture seems set to gain momentum as pressure builds for more local production and consumption networks. In this field we highlight experiences on a number of pioneering biodynamic farms. The final chapter reviews the dramatic nature of current circumstances, highlighting concerns about food cost and security. It discusses issues which confront the organic movement and the challenge for biodynamics to be more flexible if it is to gain impetus. It points to the need for new relationships between society, the land and planet earth, all of which are inspired by spiritual ideals.

As the book's chapters are written with a progression of ideas in mind there will be some benefit reading it that way. However, they can also be regarded as separate essays, and generous cross-references are provided. It is hoped that in this way, and with the provision of numbered notes, the book will offer useful study material.

Acknowledgements

This book is dedicated to my mother, Winifred Smith, who introduced me both to gardening and to the work of Rudolf Steiner.

I offer thanks to those with whom I have been able to work and discuss biodynamics, notably to my former partner Freya Schikorr and to Matthias Guépin. These include colleagues and friends within the Biodynamic Association as well as those I have visited in the course of inspection work with Demeter UK. Friends and acquaintances overseas, notably in Sri Lanka, have helped broaden my cultural awareness.

Those who have opened doors or played a part in my work within the biodynamic movement deserve mention. I would especially acknowledge Jimmy Anderson, Pauline Anderson, Joan L. Brinch, Timothy Brink, Alan Brockman, David Clement, Anthony Kaye, Hans-Günther Kern, Manfred Klett, Walter Rudert, Patricia Thompson and Olive Whicher.

I am grateful to Dr Margaret Colquhoun and Bernard Jarman for contributing Chapters 9 and 11 respectively, while Mark Moodie and Simon Charter provided assistance with Chapter 8. Others have offered valuable comments, including Alan Brockman, Peter Brinch, Wendy Cook, Bernard Jarman, Hans-Günther Kern, Dr Nicholas Kollerstrom, Paul and Anny König, Dr William Smith and Hans Steenbergen. Finally, various suggestions for improvement have arisen during the editorial process.

1. The Foundations of Holistic Agriculture

A historical context can help readers assess the distinctive contribution that biodynamics offers for understanding the natural world. Here we shall review how people in the past approached agriculture, and what has happened as a result of using agrochemicals. Organic farming will then be considered, noting the many benefits this can bring, before finally introducing biodynamics.

The nature of indigenous knowledge

Truly great minds are never those of narrow specialists. If we go back in time before the last three hundred years we find that knowledge was drawn from a wider field than tends to be the case today. Many great minds of the past—Leonardo da Vinci, Kepler, Shakespeare—could thus be described as polymaths, for their creative strength lay in a universe of knowledge to which they were still receptive or in which they were well versed. Chaucer, for example, commented that no one could be a physician who was not also informed by astrology,[1] and when we use the word *consider* (*siderus*, Lat. star) we can appreciate that originally it meant to 'consult the stars'. Holism, as it arose in Greek times, reflected the merging of a dawning intellectual faculty and its keen observation of the world with an intuitive wisdom which had prevailed for thousands of years. The contribution of Aristotle was pivotal in this respect. The Greeks stood at a crossroads of human evolution for they recognized the reality of a realm beyond material physical existence, embodied in their mythology. Awareness of cosmic influences on earthly life, as will be discussed in Chapter 2, was common amongst all ancient peoples. As we have increasingly gained independent intellectual skills, we have largely lost a faculty which once gave us access to a universal wisdom.[2]

To understand indigenous agricultural practices, some of which offer examples relevant to our time, we need to realize that ancient peoples, in particular their priestly elites, had an intimate understanding of natural phenomena. In the mystery centres, they were able to consult with cosmic wisdom (oracles) on all aspects of cultural life, including agriculture. Such access was normally acquired via a path of initiation (Fig. 1.1). Nowadays we assume trial and error has played a key role in human progress, including plant and animal domestication. This is the

Fig. 1.1 Angelic being depicted as pollinating flowers. Assyrian bas-relief from Nimrud, Iraq (Boston Museum of Fine Arts)

brainchild of an intellectual era, for the first domestication of grasses was achieved many thousands of years ago while a broad range of evidence shows that in temple architecture, music and painting, sublime expressions of artistic endeavour preceded later and lesser achievements.

But living with nature year after year, in the rhythm of seasonal changes, was more than an education; it was the profound, first-hand experience of ancient peoples aware of the outer reality of their environment but also aware, inwardly, of how harmony or balance was to be achieved on the land (see Chapter 9). Balance is of course the aim of sustainable agriculture—farming which puts a premium on building and maintaining the quality of the land resource.

Traditional cropping systems

Until the nineteenth century all the world's agriculture embodied elements of holistic practice. In Britain this consisted of relatively modern practical measures (rotation, tillage, drainage) with an underlay of tradition (use of locally adapted species, timing of agricultural activities, observance of seasonal festivals, folk knowledge of medicinal plants). Each indigenous system adopted different approaches to building and maintaining soil fertility but from ancient times it was evident that the waste discarded from settlements—from animal housing in particular—led to prolific plant growth, so the principle of applying surface organic materials became widespread. This can still be read from the dark appearance of soils in the vicinity of ancient settlement sites.

Many systems relied on combinations of fallowing and the collection of various nitrogenous materials for incorporation into the soil. Field-based and largely rain-fed systems from Europe to Asia depended especially on the manure from draft animals. These were not available in the Americas at the time of European contact, and other materials were used here including the guano of sea birds, together with fish heads.[3] The Celtic coastlands of north-west Europe traditionally made use of kelp and other seaweeds which they laid out in long, narrow beds—'lazy beds' (Figs 1.2 and 1.3).

Fig. 1.2 Seaweed beds on the Isle of Rhum, Scottish Hebrides

Fig. 1.3 Harvesting seaweed in Brittany, France

Other systems relied on the natural fertility of fresh silt, flood and irrigation water to provide for sustainable production, the Nile, Indus and Mekong being classic examples.[4] For this and other reasons, rice growing throughout the East—often virtually a monoculture—as well as wet yam growing in the Pacific region, has been amazingly robust. For example, paddies in parts of the East would be fertilized by the leaves and branches of surrounding trees, some of which were nesting sites for fruit bats which thus delivered guano. Buffalo would contribute by treading in crop residues and other organic matter while also adding dung. Meanwhile, after irrigation or flooding, paddies would benefit from blue-green algae, and even fish! Not so any longer, in the vast majority of areas.

In forested regions the food production strategy depended on whether native species offered enough subsistence or trading potential to be worth exploiting. In this way, and subject to other land being available for field crops, forest gardens came about. Much advocated by permaculturists, these are highly productive systems but mainly appropriate for tropical environments. Such sophisticated forms of polyculture are thus still practised locally in Sri Lanka (known as Kandyan gardens), parts of Indonesia and other areas of fast-diminishing rain forest. Here, the forest provides all its own nutrient needs while the diversity of species and ecological niches create a vast food web within the ecosystem. This is the eminently suitable form of productive land use for sloping terrain, and many of the best examples owe their survival to the difficulty of other types of cropping on such land. Even so, cultivators formerly constructed terraces to maximize food cropping and conserve soils against erosion.

For the growing of field crops, farmers widely used the system of 'slash-and-burn' land preparation, given names such as swidden, ladang, milpa, tlacolol and chena in different parts of the world. It is also thought that the brief burn of the soil surface helped stabilize nitrogen until the next rains came and seeds could be sown. After several seasons of cropping, the land was left as fallow for a lengthy period. Such systems achieved a reasonable equilibrium as long as land resources were plentiful in relation to population levels. During the cropping years it was common for different main crops to be planted in succession, both perennial and seasonal, according to their nutrient demands. Eventual abandonment, while allowing the bush to return and fertility to be restored, reflected the extra work involved as weeds became more troublesome. Especially in lower latitudes, cropping utilized not one but several species in combination, for example maize, beans and a cucurbit (Mexico and Peru), or alternatively manioc with either pigeon pea and sweet potato (Indonesia) or cow pea and melon (Cameroon). Such polyculture was worldwide prior to the advent of chemicals and modern farming methods—in fact prior to modern agricultural science!

It has been repeatedly shown that crops grown in combination outperform those grown separately over an equivalent area (the Land Equivalent Ratio), and that for poorer soils the benefits are proportionally greater.[5] We should not be surprised by this, for the net primary productivity of complex ecosystems such as native forests is as much as two orders of magnitude greater than for modern arable farming. The benefits from mixed cropping are not trivial either, for increases of 1.5 to 4 times have been recorded in experiments. This arises from efficient utilization of soil and light, beneficial interactions between neighbouring plants above and below ground,[6] effective pest management, and from comprehensive ground cover which stifles weed germination and helps maintain soil moisture. Also implicit in this system is a spread of the ripening and harvesting of each component. Polyculture was fundamental to food security, for if one crop failed, the family might survive on what remained. In Japan, barley and sweet potato used to be interplanted with *Calendula*. Intercrop systems were also widespread in China, though recent reports indicate those still adopting such practices are reluctant to talk about them as they tend to signify poverty. In fact, however, nothing could better exemplify holism in practice than such a cropping system. But with increased scale and mechanization, diverse cropping systems have become impractical. In many parts of the world, principally the tropics, we can see the remnants of such ancient systems. Even where mainstream agriculture has erased them they survive in home gardens where cultivation is based on bed systems and hand tools.

In higher latitudes with a lower angle of sunlight and more limited growing seasons, simpler cropping systems—whether in beds or fields—would have been more practical. The keeping of animals also had an effect on how cropping could be organized. In Britain, a strategy involving fallowing—swidden in prehistoric times—had been the norm. The medieval-village 'three-field' system embodied this idea and incorporated so-called 'open-field strips'—very much an expression of people still labouring communally. But from Tudor times these strips were consolidated, a process accelerated by Acts of Parliament, and largely driven by the Industrial Revolution. To increase the intensity of production yet avoid soil exhaustion, cereals were followed by other crops—roots and later legumes—giving rise to locally distinctive rotation systems. Some idea of the earlier character of agriculture and biodiversity throughout Europe can be gleaned from recent studies of Transylvania.[7]

Even so, European single-variety rotations were as nothing compared to those of Central and South America or even West Africa. In the Andes, fields nominally planted to maize or potatoes in different seasons are commonly interplanted with other species, while not just one variety but up to 30 varieties of the main crop can be identified! If this is not a measure of our agronomic regression in recent times then it is certainly a measure of the genetic erosion which modern practices have caused.

Traditions of sowing and planting

There are countless local traditions for sowing, planting and other agricultural practices across the world. Extant in China in the Middle Ages was an elaborate calendrical system incorporating moon and constellations (Fig. 1.4)[8] while a system of *neketh* or timings is still used in India and Sri Lanka.[9] This not only concerns auspicious timings for work on different crops but, through the use of *tithi* (denominations of the lunar phase

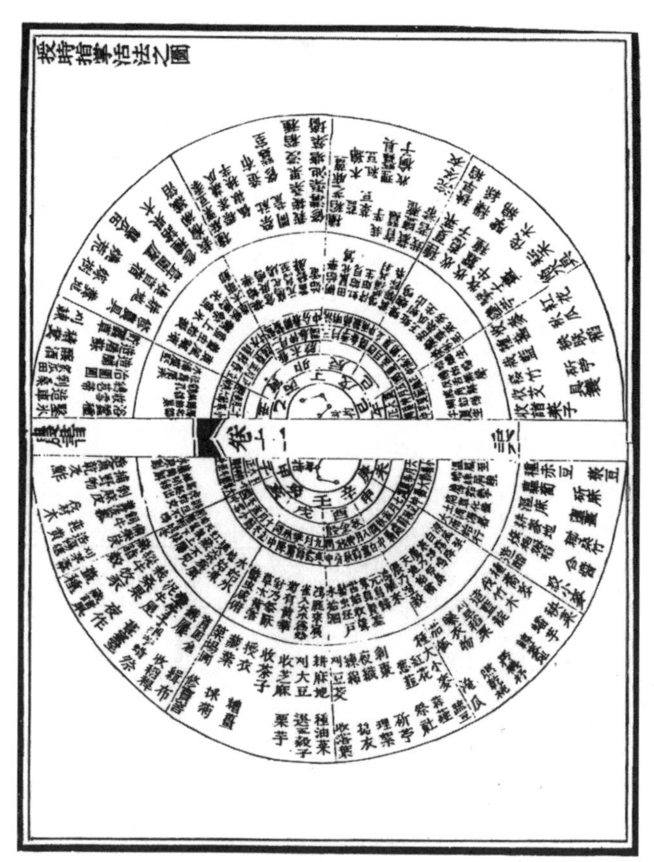

Fig. 1.4 Calendrical diagram from the Wang Chen Nung Shu, *China,* AD *1313 (from Francesca Bray, 1975)*

cycle), optimal timing can be assigned for individuals according to their horoscope!

In the tropics where, rainfall permitting, cropping is possible throughout the year, the moon's phases feature strongly in indigenous knowledge. Seasonal positions of sunrise as well as lunar rhythms used also to be keenly observed in temperate regions. Among many variations, the writer has detected similarities with regard to sowing and planting. Thus, traditions worldwide tell of *sowing* 2–3 days *before* the full moon and of *planting or transplanting* 3–4 days *after* full moon. It is acknowledged that the moon strongly influences water processes while seeds absorb water more rapidly around full moon (Fig. 1.5).[10] Germination is therefore more effective if seeds are sown at that time.

One of the priorities of farming is to achieve rapid and even emergence of seedlings, so besides attention to seedbed preparation, seed priming may be undertaken (see Chapter 7). Without this, many seeds are victim to fungus or predation, while slow emergence invites weed competition. Agriculturalists of the past appear to have anticipated the optimum water uptake of seeds around full moon by sowing a few days before. For transplanting from a nursery the situation is different, for here we have already a plant with its root system. In this case, re-establishment of roots in fresh soil is the priority so that planting out in the moon's waning period—when water is drawn less strongly within the plant—is entirely prudent.

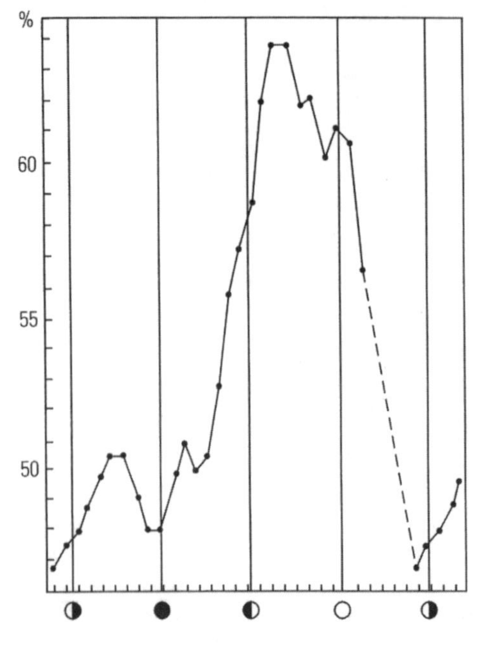

Fig. 1.5 Water absorption by bean seeds around full moon (after Brown and Chow, 1973)

Decisions about what and when to plant were made also on the basis of keen observation. Indigenous knowledge systems thus included awareness of the habits of a range of wildlife species as portents of the weather in the seasons ahead.

Traditional systems of pest management

Traditional pest management took different forms. There were cropping combinations which helped contain such problems, together with attempts to scare pests or lure their predators by day or night using a variety of ingenious devices (Fig. 1.6).[11] Auspicious timings were used to control different combinations of animals. Thus the phase cycle of the moon was divided into seven-day sub-cycles, each day deemed to have a particular characteristic 'energy'. Such days, accorded the names of animals, are known as *karana* in Indian and Sri Lankan tradition.

The shaman's services were widely used for tasks seeking control over nature that could not be undertaken without attaining high moral standing. Mantras would be chanted—stanzas which had power when recited with correct emphasis and repeated a particular number of times. Such rhythmic utterances are aimed at influencing the powers guiding the activities of different animal groups (group souls) and the world of elemental spirits (see below). In a similar way, *Pirith* ceremonies in Buddhism, which make use of water, offer protection for the person, for properties and land. But such mystical procedures, as with the Agni Hotra *puja* (offering),[12] may also involve a burning process (dematerialization), a reflection of which is found in the church's use of incense. While shamanism has always been concerned with human welfare (e.g. the

Fig. 1.6 Indigenous devices for pest control, Sri Lanka. Left: Attracting birds to clear ground-dwelling pests. Right: Nocturnal lure for insect pests

Fig. 1.7 Shaman invoking powers to protect potato field, Bolivian Andes

witch doctor), it addresses not only the hazards of pests and disease but, in the Andes, protection of potato fields against violent hailstorms (Fig. 1.7).

Meanwhile a traditional method by which gamekeepers controlled the destructive habits of crows was to kill a few and place them on fence posts to deter their brethren, and some other pests are treated in a similar way to this day in parts of Britain. In a similar fashion, rice insect pests in paddy fields in Sri Lanka were once widely deterred by pinning said insects on posts at the four corners of fields. The choice of particular times or of chanting reinforced the deterrent effect (see Chapter 5).

Traditional reverence

It will seem strange to western people—now almost completely separated from the land—for this subject to be brought up in the context of agriculture. The sanctity of the harvest, together with sowing and other seasonal events, has been and continues to be a feature of traditional societies and, though poor, many in the developing world spend heavily on festivals and other celebrations. Formerly, in Britain and elsewhere, observance of each lunar cycle according to a system of Esbats and Sabbats was once widespread.

From the various theocratic religions of Asia and America to the animistic traditions of Africa, deities and elemental nature beings have traditionally been acknowledged. In these traditions there is recognition it is not only human work which gives us our crops but the participation of unseen helpers and unseen forces, now largely beyond human experience. Before starting work, many Hindu farmers still worship at a shrine placed within the field (Fig. 1.8), while in other parts of the world a consciousness of nature spirits remains. On a village development programme in Ghana,

Fig. 1.8 Hindu shrine beside rice paddies in Bali, Indonesia

the writer encountered people who could engage in such communication. He was informed that these spirits were now departing because of the noise and pollution of tractors and the clearance of 'bush', which left them with no natural places to which they could retreat. This is no isolated incident, for recent sources illustrate the reality of a highly structured body of knowledge about nature beings.[13] Many people have a capacity to perceive such spirit beings, more so when they are children, but such experiences are, of course, completely at odds with western paradigms.

Modern science assumes that with similar techniques different people can achieve similar results—but we know this is simply not the case. Individuals most certainly do have an impact on the world of the living; and a reverence for nature, whether displayed or not, appears to be fundamental to this (see Chapters 3, 4 and 5). We may therefore begin to understand how people in the past, and perhaps not just the shaman, were able to exercise influence over some of agriculture's natural hazards.

Chemicals and commercialization

The history of conventional, chemically based agriculture parallels that of developments in science, warfare and commercial activity since the early nineteenth century. The details of this are outside the scope of the present volume but the lessons are not.[14] We may be reminded here of the contrast between *need* and *greed* as famously pronounced by Gandhi. The advent of chemical fertilizers meant that farmers were no longer dependant on cultural strategies to maintain nutrition for crop growing. It simplified the farmer's life but was also to be the start of rural indebtedness. Chemicals led directly to continuous cropping, to a decline in soil organic matter, soil organisms and soil structure, with urea a key culprit.

A further step was made when Norman Borlaug's 'miracle' seeds hit the market place in the 1960s. These were the high-yielding, hybrid cultivars which were to give the Green Revolution its name. And chemicals, of course, were a necessary part of living that particular dream. They were needed for growing the crops to their potential yield, for preventing pest damage and for reducing weed competition. How ironic that when the chemist and the business world uses the word 'green' it usually means the exact opposite!

In consequence, a small number of hybrid varieties displaced countless open-pollinated traditional cultivars (see Chapter 7). This loss of biodiversity spelled the beginning of a continuing campaign against pests and disease which fails to be won despite huge expenditure. Enlargement of fields and farm units in the course of time to facilitate mechanization has merely increased these problems. The uniformity which resulted from the new varieties placed crops at risk because fewer organisms find suitable niches. Short-strawed cereals help avoid wind damage to heavier crops but bring the crop nearer to the ground, reducing ventilation and increasing the incidence of fungal disease. The introduction of genetically engineered crops is but a further turn of the screw which, despite current belief in technology, will not prevent future plagues of pests.

This narrowing of the range of crops has also meant that part of the broad nutritional base has been lost from local diets the world over. Thus a great deal of malnutrition is directly attributed to changes following the Green Revolution—an increased incidence of children with rickets being one example. Looking back to the nineteenth century, the lack of agricultural diversity was the key factor in the infamous Irish potato famine.

It could be claimed that agrochemicals have helped the world's population climb from 1.5 to over 6.5 billion in the last hundred years. On the other hand, the environmental and human costs of achieving this have been enormous: soil erosion and desertification, water pollution, greatly reduced biodiversity and health risks to farmers and consumers alike—all this before considering the implications of increased carbon dioxide and other greenhouse gases, a process to which an intensified agriculture has made its own sizeable contribution.

Pesticides

The story of pesticides is a story of self-defeating aggression. It is also a story of bad science—bad because of the adverse consequences of pesticide use, and bad because most of the research has been conducted under the control of those with commercial interests. Pests of field crops usually have control organisms. When pesticides are applied, their natural enemies are killed along with the target organisms. Spraying affects the entire food chain dependant on insect life or other food sources.

Susceptibility of plants to pest and disease attack is crucially affected by stress. Arising from many sources, stress causes changes in metabolism. For example, prolonged intense solar radiation causes stress, so that temperate C3 plants are more prone to insect attack in the tropics. The provision of shade is therefore a vital consideration. Stress can also be caused by pesticides interfering with the surface protective layer of the leaf, then being absorbed by the plant. While plants can to an extent metabolize pesticides or their fragments, the latter can lead to retarded plant metabolism. The resulting elevated sugar levels may increase susceptibility to insect attack.[15] The latter not only feed on the plant but also introduce virus diseases. In addition, pesticides inhibit soil micro-organisms such as nodule bacteria and mycorrhizal fungi, so for this reason too plants may struggle to maintain healthy growth.

Owing to their rapid and frequent life cycles, pest organisms are well placed to develop resistance to the various chemicals—this principle also applying to virulent weeds. The result is that the farmer either sprays more frequently or increases the dose, irrespective of what the small print on the bottle may say. This intensifies the evolutionary pressure to acquire resistance, as well as increasing the health risks to the farmer. Pesticides and herbicides are formulated with a carrier substance which allows them to stick to or penetrate biological tissue more readily. While most of this propellant-surfactant is evaporated quite quickly according to humidity, it is not uncommon for the chemically active agent itself to have volatilized within 48 hours. This highlights both the pollution risk and the comparatively short-term effectiveness of many pesticides. Indeed, it has been estimated that at least 80 per cent of pesticide is wasted,[16] and due to the world's air movements temperate regions receive an undue share of the deposition of chemicals carried upwards in the atmosphere of the tropics.

Since the Industrial Revolution and major urbanization, living in the countryside has generally been healthier than town life with its pollution. But since the advent of agrochemicals, farmers have been at particular risk from pesticides, particularly poor farmers in hot countries who work unprotected and without access to adequate water for washing.

Organic-ecological farming

Without the recent history of so-called 'conventional' chemical farming it would be unnecessary to label any form of agriculture as *organic* in a generic sense. The word organic (or *biological, ecological* in other languages) signifies farming according to *standards* or agreed sets of principles. This sort of farming is 'non-chemical' in the sense that it does not use synthetically produced fertilizers or pesticides, nor does it permit the use of genetically engineered crops or products. And in order to ensure as far as possible that only genuine organic products reach the marketplace, each producer, processor, packer or importer of organic produce is subject to annual inspection in order to maintain what is referred to as *certification*. So the term 'organic' indicates a type of production but also signals customer assurance in the marketplace. Thus no produce may be labelled 'organic' and receive an organic logo unless it has been through such a process.

We will not attempt a historical review of this kind of agriculture, as excellent coverage already exists.[17] A common misconception is to pigeon-hole it as merely 'traditional'. Detractors often choose this direction of attack, for they like to promote the idea of organics as unprogressive and backward-looking, even 'medieval'. But agriculture has a long heritage so it is natural and healthy that ideas which have proved themselves successful will live on into the future and become adapted in the course of time. In this way, different schools of thought originating from twentieth-century pioneers have contributed to a range of recognized organic practices. Additionally there is Permaculture, a complementary movement driven by a design philosophy. This is concerned with sustainable living as much as agro-ecological strategies appropriate for different local environments.[18] The latter actively supports multiple-crop strategies and agro-forestry.

Differences between organic and conventional (non-organic) agriculture

What then, are the main differences between chemical and organic agriculture? In a nutshell, conventional *chemical farming* supplies nutrient for direct use by crops—soil is the medium in which this takes place. Nutrient concentrations fluctuate widely in the course of the year, influencing the behaviour of plant roots and micro-organisms. Wastage of chemicals occurs—especially under tropical conditions. The principal approach to pests is to target offending organisms—in other words to treat symptoms rather than causes. Environmental and human costs are externalized in the drive to maximize productivity.

For *organic farming*, building and maintaining a healthy and fertile soil is fundamental and can be illustrated from the sample data shown below where the 'organic' soils have higher water-holding capacity, a less compact nature and, through better rooting and biological activity, they extend to greater depth before the parent material is reached. The driver for all this is a combination of rotated and mixed cropping, green manures

Mean values of soil properties from organic and conventional farms (after Reganold *et al.* 1987)

Property	Organic farms	Conventional farms
Surface soil colour (Munsell)	10YR4/2	10YR5/2
Polysaccharides (g kg^{-1})	1.13	1.0
Moisture %	15.49	8.98
Surface texture	Silt Loam	Silt Loam
Bulk density (mg m^{-3})	0.98	0.95
Modulus of rupture (MPa × 10^{-2})	1.61	1.98
Surface org. horizon depth (cm)	39.8	36.7
Depth to compact subsoil (cm)	55.6	39.8

and the use of recycled organic wastes. In the majority of cases animal husbandry will also play its part. Organic farming and gardening provides plants with access to nutrients delivered by slower-release soil processes. The primary approach to pests is to use methods which promote a diversity of organisms. The aim is to produce a sufficient supply of healthy food while preserving the environment for future generations.

As part of rural economic life, agriculture has always sought to achieve

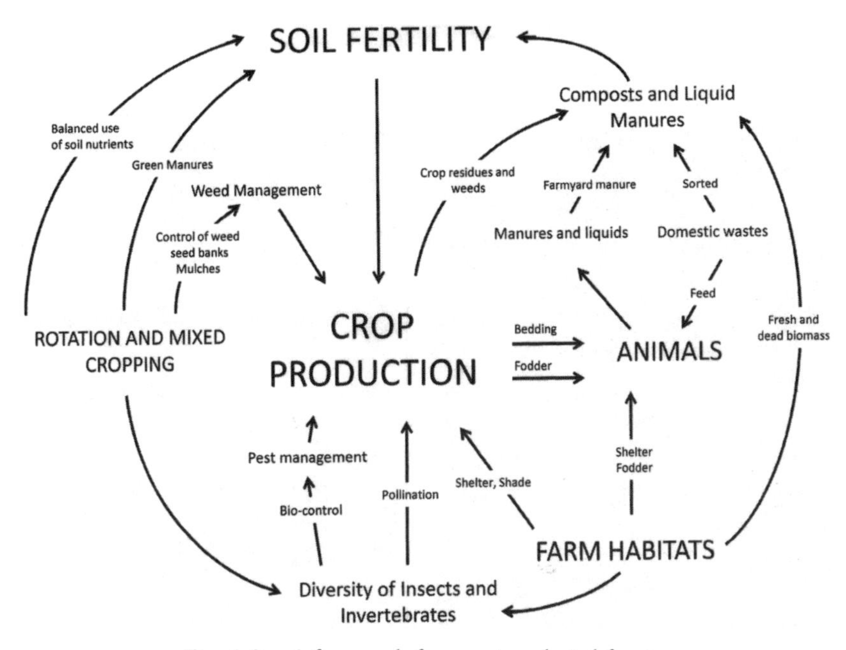

Fig. 1.9 A framework for organic-ecological farming

two things: to provide a suitable basis of fertility, and to limit the losses due to pests. These twin realities are recognized in Fig. 1.9, where it is evident that organic farming cannot be viable without also being *ecologically based*. There is a parallel with the human being, for while we require nourishment (inputs) we also depend on an immune system (defence). Indeed, the essential philosophical difference between chemical and organic farming is that the latter formally recognizes the connectedness of things. It is for this reason that the ideal organic farm has been likened to an organism (Chapter 3). Organic farming must therefore go beyond a mere substitution of chemicals by organic fertilizer inputs. Indeed, its approach to plant nutrition and control of pests illustrates a potential robustness which deserves further comment.

Building soil fertility

Figure 1.9 shows a certain focus on the making of compost—manures and compost continuing to mark out organic farming and horticulture in people's minds.[19] Nevertheless, it is impossible to rely on compost as an input, so soil fertility is normally maintained in the rotation by growing green crops which lift nutrients and build nitrogen reserves. Some operations may depend heavily on the latter. For reasons to be elaborated in Chapter 3 this is far from ideal, but it can be a route to achieving organic production over large areas. Because the majority of organic farms and most home gardens will be handling compost, further comments are now offered (Fig. 1.10).

Fig. 1.10 Compost making as part of the educational process, Sri Lanka

Here we have to consider the most beneficial and least wasteful way of returning organic materials to the soil. To apply manure or slurry to the surface of land can be wasteful, although every method has its correct application and slurry applied to pastures as early grass growth occurs is a tried and tested practice. But if possible, organic materials should be composted, for in this way the nitrogen is more stable, especially under a regime of leaching, while the soil will not be so prone to developing acidity. Composted manure applied in the autumn to the soil surface will generally be carried down into the soil by earthworms. Given the tendency for increasingly mild and wet winters in the UK it is more important than ever for organic farmers to manage their nutrient resources wisely.

So what actually is compost, why is it so valuable to us and why therefore should organic practitioners take trouble to make it in a professional manner? Practically speaking, compost contains a high proportion of humus, which is what all organic matter breaks down to in the soil. The problem is that in the course of breaking down in the soil there is likely to be a shortage of nitrogen for plant growth depending on the carbon to nitrogen ratio of organic matter added. A soil environment in which decomposition processes predominate conflicts with healthy root development and crop establishment while a soil containing high levels of humus colloids is of the greatest benefit to plant rooting (Fig. 1.11).

Compost can be made from many different types of plant and animal waste and depends on the bacterial breakdown of carbohydrate- and protein-based substances together with incorporated minerals (see also Chapter 3). Providing there is adequate aeration, the organic matter is broken down initially to ammonia which dissolves in the moisture present to form ammonium ions. These in turn are oxidized progressively to nitrite and nitrate by different bacterial strains. It is in the form of nitrate that most plants take up nitrogen for subsequent elaboration of amino acids and proteins.

Besides nitrogen, completed compost contains sulphur, phosphorus and all the major and minor nutrient elements. For example, all green matter consists of chlorophyll which incorporates magnesium as well as nitrogen. Composts will vary in their chemical composition but in general they provide a well-balanced range of nutrients for plant growth. The organic matter itself is of benefit to the soil's physical consistency, contributing to the soil structure and water-holding capacity. Compost is both a source of, and substrate for, soil micro-organisms which are the essence of a healthy soil. The added humus also improves a soil's capacity to hold nutrients, a property known as the cation exchange capacity

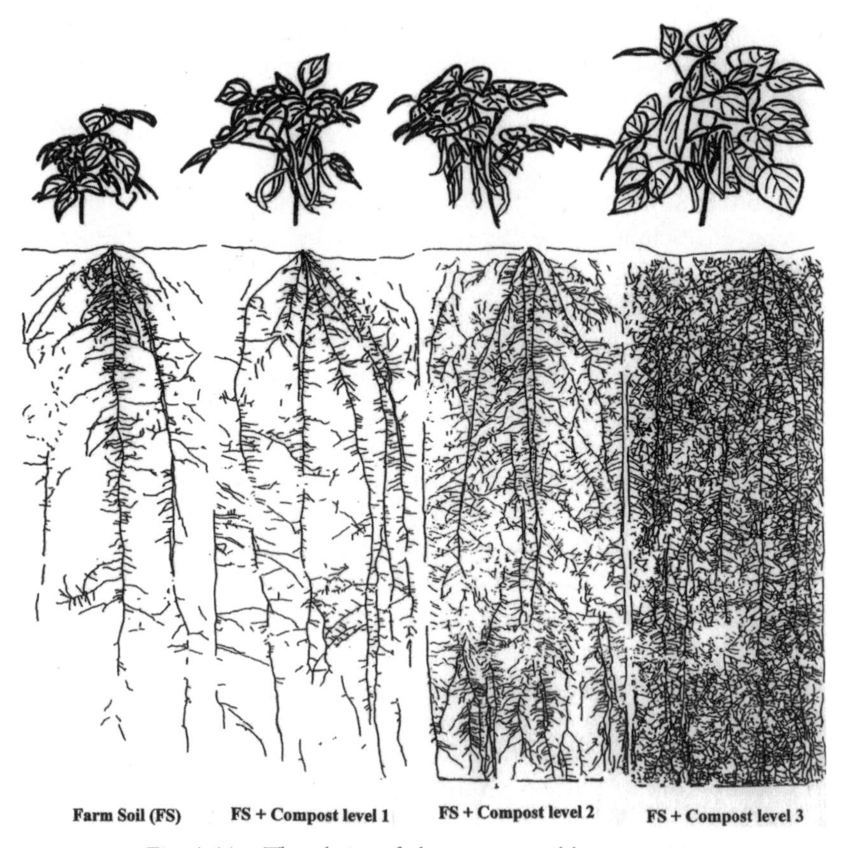

Fig. 1.11 The relation of plant roots to soil humus content (from Jochen Bochemuhl, 1981)

(CEC). In many soils a comparatively small addition of compost, providing this can be sustained year-on-year, will significantly enlarge the CEC (see also Chapter 4). In tropical soils, this can have spectacular results and is, more demonstrably than in cooler countries, a foundation for the success of a sustainable organic agriculture.

Management of pests and disease

The main object of organic farming is to build health rather than fight disease and pests. If we think about plants growing in the wild, the issue of disease and pest attack rarely arises because in addition to wild plants being more robust, the diversity inherent in the plant community effectively controls any organism getting out of hand. The message is that we have to design our agricultural ecosystems—on both larger and smaller scales—to be as species-diverse as possible.

As portrayed in Fig. 1.9, diversity of insects and invertebrates is a cornerstone of pest management and there are several ways in which this can be achieved. Creating and maintaining habitats for wildlife, such as woodland and hedges, together with ponds and wetland areas, is essential, using species native to the area as these are usually host to far more insects than introduced plants such as ornamentals. Both arable rotation and mixed cropping contribute to the management of pests for they signal plant diversity, and whether this is generated in space or time, diversity of insects and invertebrates will follow (Figs 1.12 and 1.13).[20] The latter includes organisms or their larval stages in the soil. Changing the crop means changing the characteristics of the surface litter layer and the rooting environment (rhizosphere) so that certain organisms—parasitic nematodes for example—cannot go on multiplying, while the numbers of their cysts will diminish through predation. So the aim must be to provide an environment in which organisms generate mutual control. Any particular outbreaks of pest and disease may then be contained more easily by the use of a range of environmentally friendly methods.[21] These include the application of a herb extract such as that based on nettle, the use of compost teas and use of imported materials such as neem (azadoractin). Biological control can also be achieved by imported pathogenic organisms such as *Bacillus thuringiensis* and different types of insect predators.

When considering pest susceptibility one also needs to consider a

Fig. 1.12 Intercropping in New Zealand, North Island

Fig. 1.13 Intercropping at Oaklands Park, Gloucestershire, UK

plant's natural defences. It is standard practice for farmers to select crops for their pest and disease resistance. In this respect, indigenous cultivars and primitive varieties are sometimes less susceptible, or even resistant. As plant domestication has proceeded, in order to enlarge the organs we wish to harvest, other attributes have become weakened (see Chapter 7). In this way natural plant defence through chemical inhibition (allelopathy) has been progressively reduced. Even so, the cells of organically grown plants have higher levels of accessory plant defence substances, the same ones we associate with flavour and, in particular, bitterness. These include phenolics, cyanogenic glucosides, tocopherols, beta-carotene and flavonoids. Organically grown plants also have thicker cell walls than those grown with synthetic chemicals.[22]

We should also address what it is that makes plants prone to pests and disease in the first place. Earlier we referred to various stresses. Plant health arises from a balance among the various processes involved in the vegetative and ripening stages of crops. Under a well-operated organic system, because nutrients are released more slowly by biological processes, such a balance is achieved and plants are less susceptible to pest problems. If we consider a community of wild plants, inter-specific competition is the norm and at no time could one plant ever experience luxury uptake of available nutrient. It is undoubtedly flushes of nutrients into plant tops which promote pest problems. One can observe that the type of growth with chemicals is different from that with organics—a rich green lushness often characterizes the former. Growth rates and yields are commonly reported as higher using chemicals—though in temperate rather than tropical experience. Yet it makes little sense boosting plant growth if the result is increased pest attack, post-harvest losses and a polluted environment.

It is therefore important to understand that *fertility inputs and control of pests are not separate issues but interrelated.*

The benefits of organic farming

Before leaving the better-known territory of organic farming and gardening it is important to underline the advantages arising from such husbandry as compared with chemical methods. These are summarized below but it is of greater value to appreciate the connectedness of the different aspects involved. This is portrayed in Fig. 1.14.

Produce quality and health. Organic food is recognized to be beneficial for health and is recommended in a number of therapies including those for allergy sufferers. Organic and less intensive methods lead to reduction in the incidence of veterinary problems and associated costs. Compared with untreated conventional produce, fresh organic products have superior keeping quality.

Soils and organic matter. Organic matter reserves are increased under organic management. This results from better plant rooting as well as from cultural practices including mulching, green manuring, compost application and appropriate cultivation practices within a rotation system. Improved soil structure results from this, giving better soil stability under rainfall impact. While carbon fluxes are an inevitable part of soil processes, organic units may claim to be working effectively towards carbon sequestration.

Plant nutrition. Access to nutrients is mainly through biological release. Organic matter levels normally ensure adequate nitrogen and other nutrients while the resulting humus enhances CEC. The activity of nodule bacteria and mycorrhizae is increased, which improves access to micro-nutrients. Phosphate fixation is countered by the flux of organic anionic compounds.

Drought tolerance. Organically grown plants have thicker cell walls which increase resistance to wilting. Deeper rooting and higher levels of soil organic matter improve access to soil moisture. Roots with highly

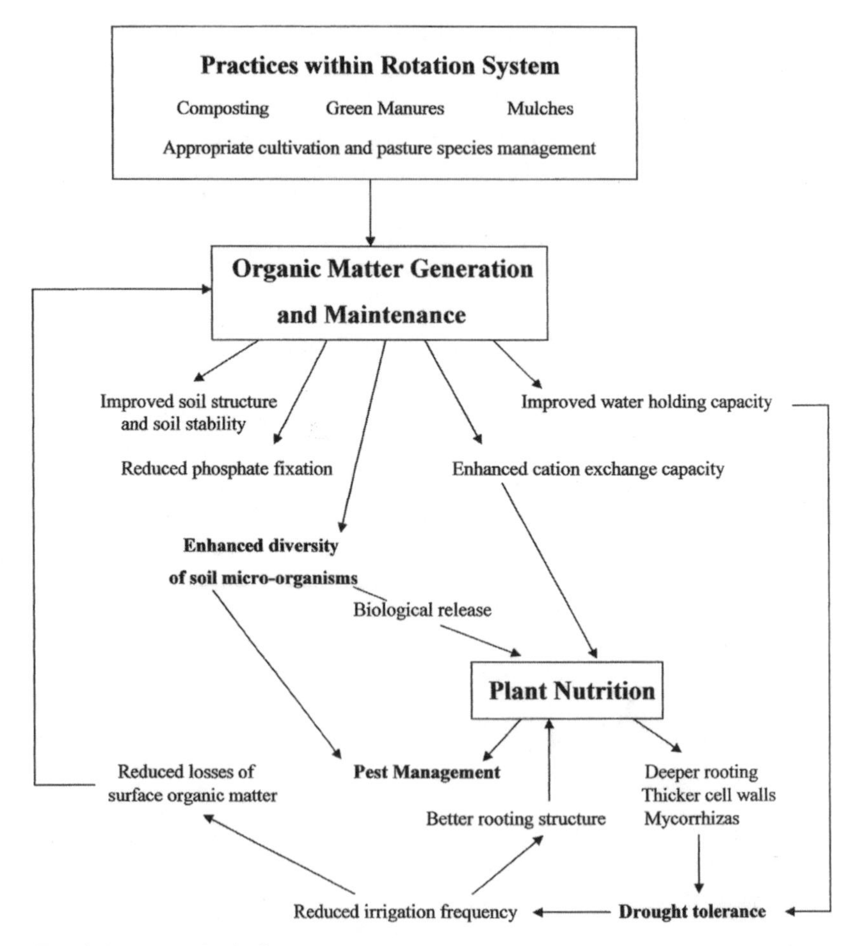

Fig. 1.14 Beneficial effects and interactions of organic farming and gardening practices

developed mycorrhizal systems have increased drought resistance. As a result, irrigation frequency may be reduced which in turn reinforces strong rooting. Reduced wetting and drying in the topsoil reduces losses of soil organic carbon through microbial metabolism.

Water. There is little risk of contamination of surface drainage waters or ground water by nitrate or other soluble materials providing compost areas are kept covered and animal yards are equipped with suitable liquid storage. As water quality and availability become ever more critical, organic farming will play a positive role in our agricultural future.

Energy. Organic farming has a significantly lower 'carbon footprint' than conventional agriculture due to the latter's consumption of energy for agrochemical production and transportation. As energy reduction plans gain momentum, organic farming is well placed for using predominantly local resources.

Wildlife. A variety of studies show that organic farming leads to increases in the numbers and diversity of wildlife. Together with a diversity of farm activities, this makes organic farms attractive for educational visits and community involvement in agriculture.

Biodynamics

In view of the formidable list of benefits for soils, environment and health arising from the adoption of organic practices, one might reasonably ask whether biodynamics can offer anything more.

The first thing to say is that biodynamics is founded upon good organic farming and gardening principles and is not something apart from it or even a substitute for it. At the risk of sounding over-simplistic at this point, it involves working more deeply with nature's processes and, in so doing, striving for produce of the very highest quality to promote the health of human beings. The organic farming movement deserves credit for bringing a healthier and more ecological approach to land work and consumer consciousness. Biodynamics carries this picture further, in particular, by addressing a system of energies underlying life processes. One can rightly say that the realm of knowledge which it uncovers provides the deeper *raison d'être* for the whole organic movement.[23]

In order to develop an understanding of biodynamics it is helpful to know why this name came into use. 'Bio' signifies *life*, from the Greek *bios* (French *biologique*; German *biologisch*), while 'dynamics' derives from the Greek *dunamikos* (French *dynamique*; German *dynamisch*), a *force* or impulse stimulating change. So when we use the word 'biodynamics' as Ehrenfried Pfeiffer first did[24] there is a presumption that we are dealing not simply with the visible forms of nature or of agriculture, but with underlying forces or energies. These 'life forces' create and vitalize nature's forms—from the germinating seed to the developing mammalian embryo.

Although the biodynamic movement began in Europe following Rudolf Steiner's seminal lectures of 1924,[25] the pressing issues of soil and health leading to it were similar to those we associate with other, if slightly later, pioneers such as Eve Balfour.[26] Steiner's lectures quite clearly dealt with a form of agriculture which could be called 'organic' but, as a reading of them reveals, they were also designed to open windows on a new way of thinking about plant and animal nutrition (Fig. 1.15). Hence it was not

Fig. 1.15 Rudolf Steiner c. *1920*

just a case of substituting recycled organic matter for chemical fertilizers, but of understanding how—via the soil—we can best facilitate the working of unseen, sustaining forces. By knowing the existence of these forces we would be able to optimize the health of plants and animals. In this regard, chemical agriculture was headed in entirely the wrong direction.

But there was something more which lay behind Steiner's motivation to speak on agriculture. He is on record as suggesting that a time would come when it would be very difficult for us to grow our crops owing to the declining vitality of the earth. We are free to interpret this as we like in relation to the world we currently inhabit, but this terrifying vision was a contributing factor to his instructions for a number of special 'preparations'. The task of these was to strengthen connections between plant life and cosmic forces (also known as ethers) and to support the essential activities of what Steiner called 'elemental beings'.[27] It is not difficult to see ways in which the earth's vitality has declined or been significantly compromised since Steiner foresaw this problem.

Although biodynamics encompasses a broad vision of the farm, of the human being and the way we work with nature, it tends to be associated in the public mind with two practical aspects:

1. To make soil and plant life more receptive (or responsive) to cosmic influences it utilizes field and compost preparations as directed by Steiner. Such measures are a basic requirement of all who work in a biodynamic way and who aspire to biodynamic (Demeter) certification. These are discussed in Chapter 5.
2. To access appropriate cosmic timings for production of different crops it makes use of a biodynamic calendar. This is largely a development since Steiner's time and is not a requirement for those seeking Demeter certification. While timings can optimize certain effects,

weather and other circumstances will frequently pose unreasonable constraints. The use of a calendar is discussed in Chapter 6.

A worldwide movement

Biodynamics has been adopted as a method of agriculture and cultivation in at least 40 countries, by many people and organizations around the world—originally largely by those inspired by Steiner's ideas in a range of other fields.[28]

With the development of a genuine organic sector within agriculture a steady proportion have chosen biodynamics as a way of achieving a distinctive quality product, as for example in viticulture and tea growing.[29] In Germany, Austria and Switzerland, biodynamics is formally recognized by government agriculture departments and research institutes for its contribution to ecological agriculture. Biodynamics has an impressive record of bringing back into production areas abandoned after years of chemical management—and this trend is set to continue. Thus large cattle and cereal farms in Australia became viable, and animals healthy, after years of decline under conventional farming methods. It is interesting to note that here and in New Zealand, isolated from its original roots, the method has been pursued with independent-mindedness and to great visible effect. This was originally thanks to the strong individual initiatives of Alex Podolinsky and Peter Proctor.[30] The latter, in conjunction with other advisers, has also played a significant part in bringing biodynamics to India. Meanwhile, a truly remarkable story is that of Sekem—a sustainable biodynamic community in the Egyptian desert, established by Ibrahim Abouleish.[31] Interest in biodynamics has also been shown in other eastern countries such as Sri Lanka, Thailand and the Philippines. Here again, it has offered a much needed land-restoring agriculture within a holistic framework which addresses traditional spiritual values.[32] Cosmic influences are no alien intrusion into life for many in these lands, as the use of a biodynamic calendar connects with the observance of auspicious timings for many important social activities. And while group certification in all these countries has provided access to export markets, it has also helped engender community consciousness.

As will therefore be evident, biodynamics connects not simply with the organic or wider environmental movements but with a holistic awareness of cosmic and spiritual influences on our lives. Concerning the latter, modern consciousness dictates that our relationship to this all-surrounding spirituality (or other-dimensionality) be based as far as possible on intellectual understanding and logical thought. It is with the latter that we shall now continue our journey.

2. The Nature of Life: Looking to the Cosmos

Biodynamics aims to understand how earth and cosmos work together. This requires us to build a knowledge of what lies outside the earth, and how it sustains what we call 'life'. So, first of all, what is life?

Life is nothing if it does not involve *process and development.* For plants this encompasses seed germination, formation of roots and leaves as part of vegetative growth, leading on to flowering and seed formation. For animals, besides growth, it includes movement and consciousness. For human beings, besides a sentient quality there is awareness of individuality, or ego. Whatever organism we would consider, the processes of development, of maturity and ageing, are all parts of living a physical existence.[1] Compared to non-living substance, living things are animated by a type of energy. As distinct from the body of physical substance which is clearly visible and embodied in the Greek *bios*, some call this energy principle a life body or etheric body—Indian culture calls it 'prana'. The Greeks, too, recognized that life had a cosmic element which infused it. This they called *zoë*. It is this which ultimately underlies all the processes of life involved in anatomy and biochemistry.

Because biodynamics is informed by Rudolf Steiner's spiritual insights,[2] our attitude towards science, the human being and evolution is considered to have direct consequences. Does one, as Sherry Wildfeuer starkly contrasts, see humanity 'as an accidental product of physical processes occurring randomly in the universe' or as a 'wisely fashioned divine creation with a capacity for love'?[3] This philosophical fault-line needs addressing before we proceed further.

Scientific method—illusions and limitations

Materialist philosophy depends on all assertions being scientifically verifiable or capable of logical or mathematical proof, and it does not accept things which are metaphysical and which cannot therefore be directly observed or measured. This is the crux of the scientific method and it defines a paradigm which has come increasingly to rule the world of the last two centuries. We may not be able to agree with this approach in its extreme form[4] but we should be ready to accept that the use of intellect and logical thinking has helped focus minds and has encouraged a spirit of disciplined enquiry in very many fields. But although philosophers and psychologists such as C.G. Jung have plumbed the depths and ascended the heights of the human mind,[5] a spiritual or religious view is mostly considered irrational—the antithesis of a scientific outlook. However, the reality of a metaphysical or spiritual counterpart to the manifest world cannot be as easily dismissed as adherents to a materialist dictum might wish.[6]

Countless people have experienced unusual states of consciousness and are convinced that a realm exists beyond the physical. 'Near death experiences' are of this kind and have a remarkable consistency of form. States of meditation are a further example of finding a way beyond solely physical perception. The inspiration of those involved with the visual arts and music offer insight here. Similarly, for very large numbers of people there is the reality of intuition which is not rational yet is capable of engendering an often trustworthy sense of certainty. Indeed, many of the greatest minds have had 'hunches'—intuitions by another name—which enabled key questions to be formulated, prior to the observations or experimentation which led to, and which took credit for 'discovery'. So if we dissect the process of scientific enquiry—if we look at it as a continuum—it is complex and cannot claim to be entirely rational. To persist in arguing that scientific and spiritual approaches are fundamentally opposed is either misguided or mischievous.

Rudolf Steiner's 'anthroposophy' (literally, 'knowledge of the human being') may be characterized as a spiritual knowledge of the human being in the context of the wider universe.[7] With roots in theosophy, it is based on his faculty of spiritual vision and its disciplined use to achieve specific research objectives. However, it is presented in such a way that the knowledge can be compiled in a manner suitable for the mind of today to grasp—hence his use of the term 'spiritual science'. In the end, spiritual research and subsequent comprehension of its findings relies on clear, logical thinking. And surprisingly, this quest for objective spiritual knowledge in no way demeans religion—it enhances our view of its role in the evolution of humanity. Most important of all is Steiner's assurance that through spiritual science we don't just exercise the intellect—we nourish the soul.[8]

Evolutionary theory—problems left unsolved

Evolutionary theory was an achievement of both logic and courage at the time it was presented. On physical evidence it appears that life has evolved through time and that higher forms have evolved from lower, like links in a chain. Human ontogenesis appears to reinforce this notion.[9] Thus 'natural selection' and 'survival of the fittest' were expressions found in the work of Charles Darwin and Alfred Russel Wallace.[10] But while the origin of *species* was addressed, the origin of *life* was not. It is this question which will remain a barrier unless certain basic ideas are understood.

Aside from the creationist view there is general belief that life arose spontaneously from inorganic substance. Yet to become an organism is to incorporate chemical substances *within a cellular framework*. The organism needs to enclose itself and become separated from an outer environment. It then needs to be capable of propagation, for which replicating signatures—DNA and chromosomes—are required. And just as photosynthesis and flowering are governed by solar radiation, there is emerging evidence that these very particular combinations of proteins are activated by *cosmic* signals.[11] Again, if we consider micro-organisms that contribute to an animal's digestive process these only begin to decompose their host when that illusive thing called life with-draws. Living organisms demonstrate an organizing principle or energy without which the body cannot be sustained. This is not just a spiritual idea but an empirical fact.

If we consider the sentient attributes of higher animals one wonders how these have evolved from purely physical origins. It is true that substances released from endocrine glands underpin fight-and-flight responses, sexual activity and so on, but what activates these processes in the first place? One might say it is our conscious reaction to circumstances, which of course depends on our life of feelings or emotions. But this is a spiritual rather than physical attribute—long recognized as the sentient or astral body, mental body or 'manokaya' (Sanskrit, *manas* = mind). Thus the astral member pervades the neurological system. How then has it evolved? And again, in human beings, where does that self-determining attribute, or ego, come from?

To perpetuate standard evolutionary theory is not to understand life or its origins. Scientific method is limited to the observation and measurement of physical reality whereas spiritual science, together with religious writings and mythology (albeit veiled and allegorical), tell of what lies behind this reality. Rudolf Steiner gave credit to what ortho-dox science had achieved but was emphatic that its methods, appro-priate for understanding the *mineral* world, would never reveal the truth about living organisms. Scientific method was able to tell us about the corpse (the geological record) rather than the living entity (the origin and nature of life). Natural and spiritual science would appear to be two sides of a coin—in our present world, one cannot exist independently of the other.

Learning from the past

At the opening of Genesis, we find: 'In the beginning the earth was without form and void, and darkness was upon the face of the deep.' Similarly, the Mayan account of creation, the Popol Vuh, says: 'Before the world was created, calm and silence were the great kings that ruled ... there was nothing. It was night; silence stood in the dark.'[12] And John's Gospel in the New Testament starts with the words: 'In the beginning was the Word, and the Word was with God, and the Word was God.' We might say that no creative process is possible without first the *idea or thought*, followed by its outer manifestation as *word and action*.[13]

Consider a meal that is about to be eaten. We may have observed the food being prepared, perhaps even purchased or harvested, but in reality the meal is a creative act originating in the mind of the person cooking. This simple example exposes the fallacy of regarding evolution as if it were inevitable. Something always lies behind what is manifest and that invariably embodies wisdom in the broadest sense. We will shortly encounter several examples of how astronomical relationships are overshadowed by mathematical principles—quite impossible to conceive if something has come about accidentally, randomly and without *thought*.

The ancient Indians and other aboriginal peoples felt God to be expressed in all the outer forms of nature. In Indonesia, the spiritual forces of the Hindu god Vishnu are brought to earth by Garuda, an eagle deity. An echo of the teachings of the Holy Rishis is experienced in later poem texts such as the Vedas and the *Bhagavadgita*. In the latter, Arjuna, a representative of humanity, engages in mystical dialogue with Krishna, a representative of the creator being.[14] This gives insight into the purpose of existence and the nature of the ultimate creative force or Godhead, for which being the name Om (or Tao) was used in the East. Today we use the Greek word *omega* meaning 'the great ultimate'. So Krishna tells us:

> The light that lives in the sun,
> Lighting all the world,
> The light of the moon,
> The light that is in fire:
> Know that light to be mine.
>
> My energy enters the earth,
> Sustaining all that lives:
> I become the moon,
> Giver of water and sap,
> To feed the plants and the trees[15]

It is evident from this original Sanskrit, as well as other sources, that great powers have been directed throughout past aeons to achieving the progress which humanity has made.[16] It is little wonder that from the ancient East through to the Druidic traditions of western Europe, profound reverence was felt for the manifest world.[17]

In ancient Asia there was a perception that life was infused by four principles: in Sanskrit, *pathavi, apo, vayo, thejo*, known later by the Greeks as those primal forces underlying different states of matter—*solids (earth), liquids (water), gases (air/light)* and an all-pervading and more rarefied principle, that of *warmth (fire).*[18] These, referred to elsewhere as ethers, were recognized as emanating from combinations of the twelve divisions of the zodiac—that circle of constellations lying along the path of the sky traversed by sun and moon (Chapter 6). In the school of Pythagoras, these primal forces and their interaction with planetary movements were experienced as musical tones, so those initiated at that time were able to speak, as did Plato, of 'the music (or harmony) of the spheres'.[19] In this connection, Rudolf Steiner explains how spiritual beings at different levels pour out their essence to make possible the conditions on which physical life is based. These beings were mentioned by name in

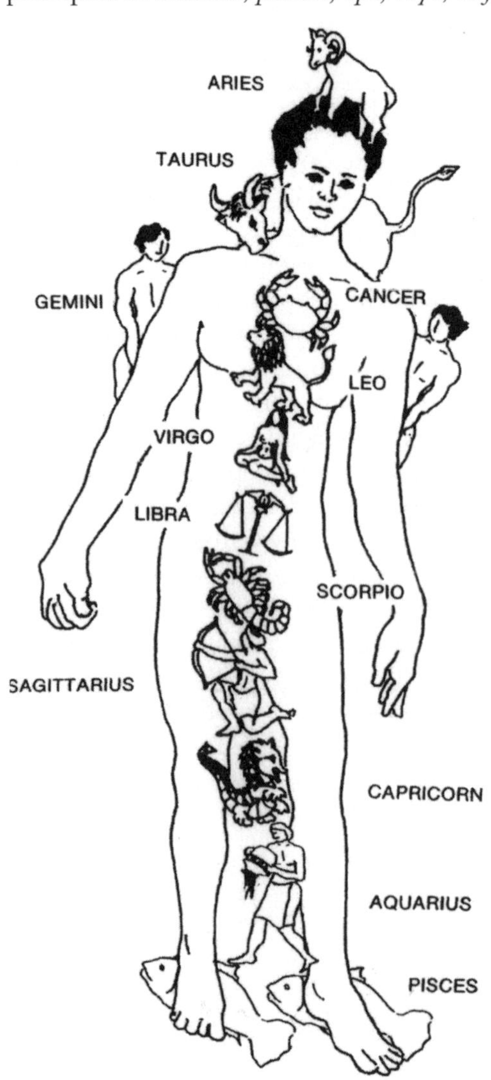

Fig. 2.1 Meganthropus *symbolizing the human being as microcosm (from Wimala Dewanarayana, 2008)*

early versions of the Bible.[20] Different groups of spiritual beings thus serve the All-Creator being, a universal spirit pervading everything from the ordered nature of the manifested 'cosmos' to the vast intervening and invisible 'chaos' of space.[21]

People in ancient times experienced the zodiacal sphere of the fixed stars as *Meganthropus*, the astrological man (Fig. 2.1), the movements of the heavens governing daily activities. We will see later that the planets were recognized as mediating these primal forces. Isolated groups of people—the Bushmen of the Kalahari, Australian Aborigines and some native Americans—still feel a connection with the cosmos. The Bushmen were surprised when the writer Laurens van der Post had slept soundly while the rest of the group had been wakeful owing to 'noise' experienced from the sky.[22] In the author's view, inspirations from such sources, though now confined to the unconscious mind, form the basis for the greater part of artistic and scientific achievement.

Among ancient peoples, the earth itself was regarded as a 'mother being'. The ancient Egyptians worshipped Isis while the Greeks experienced this female nature as Gaia. Other names were given in the course of history, including Rhea, Astarte, Hera, Aphrodite, Artemis, Pallas Athene, Natura and Demeter.[23] Druids and wicca were sensitive to this female aspect of the earth. Recent popular ecological writings such as those of James Lovelock[24] suggest we remain comfortable with this idea for, like a mother, the earth's natural systems are still—just about—able to support our physical existence. This raises the question as to whether our earth mother's ability to grow crops is merely dependant on human labour, soil and climate. Mother's fecundity would normally depend on father! So if there is a 'father principle', how does it work?

In mythology and creation accounts the principal creator beings are male; in Christianity, reference is made to 'God the Father'. The Egyptians regarded their sun god Re as masculine—the Druids, too, saw the sun in this way. Notable progenitors of culture were also *representatives of the sun*, such as Manu and Zarathustra in the Middle East, Quetzalcoatl in Mexico and Viracocha in Peru. The Hopi, indigenous Americans, still retain a cultural heritage. Their craft work includes archetypically round figurines of 'Mother Earth', while taller, sticklike characters are 'Father Sky'. Mauri tradition thinks of rain as 'tears' falling from Father Sky (Rangi) yearning for Papa, his spouse, the Earth Mother. Returning to Gaia, we find she was *wife* of Uranus, *the god of the heavens*! Evidently, there was once a more balanced picture of how earth and heavens worked together so it would be churlish to regard the Hopi crafts as mere tourist gimmicks. Further Sanskrit lines are notable in referring to the earth goddess *and* the expanses of the universe:

> O goddess Earth, O all-enduring wide expanses
> Salutation to thee.
> Now I am going to begin cultivation.
> Be pleased, O virtuous one.

So people in the past appear to have had a consciousness of supernatural beings and forces, and we know from myths and sagas that 'gods once lived amongst the human race'. Materialistic thought either considers all this to be dreamlike nonsense or requires that physically manifest beings are involved. In fact, however, we have gradually lost consciousness of spiritual worlds and been left to look after ourselves. This is the significance of *The Twilight of the Gods* in Wagner's *Ring* cycle. This separation has been necessary, for otherwise we could not have developed our present sense of individuality nor indeed our faculty of logical thinking. Loss of consciousness of the divine meant that it fell to religions to inform humanity of its origins and to provide a framework of laws for the organization of cultural life. This is the significance of the Ten Commandments of Moses which were transformed, for a later epoch, by Jesus Christ.

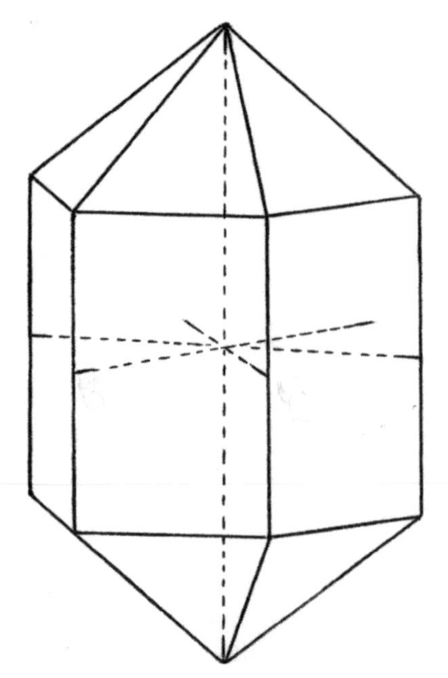

Fig. 2.2 Simple bi-pyramidal prismatic crystal of quartz. Perspective drawing using typical axial values

How can we picture cosmic forces?

Our manifest world would therefore seem to owe its existence to creative forces beyond the earth—minerals, plants, animals and even human beings connecting to archetypal influences in the universe.[25] We can attempt to illustrate these things by studying the mineral world. All physical substance represents the drawing together of forces from the periphery to a point—ultimately the atom.[26] The chemical elements, which Steiner described as 'dead images of the cosmos', are to be understood in this way (Chapter 4). Unlike the curved forms of the living world, where movement is a governing principle and where planetary influence expresses itself (see below), the defining characteristic of the mineral world is straightness and angularity of form. Crystals express in a macroscopic way the inner organization of their molecular structures. We find classes of symmetry to which different minerals belong. There is always a single axis of rotation about which two or three other axes are inclined. In the cubic system, which includes common salt and iron pyrites, all three axes are at right angles. The hexagonal or trigonal system, which includes quartz (Fig. 2.2), has three axes 60 degrees apart on the same plane, with a fourth rotational axis at right angles. So what might it be that underlies such an intersection of lines and planes?

Looking out into space we sense emptiness but 'such a thing as mathematics calls "space" simply does not exist . . . everywhere are lines of force, directions of force . . . these are not equal, they vary and are differentiated'.[27] With projective geometry we can visualize mineral structures and systems of symmetry as earthly embodiments of combinations of these cosmic forces (Fig. 2.3).[28] That seemingly lifeless minerals have their

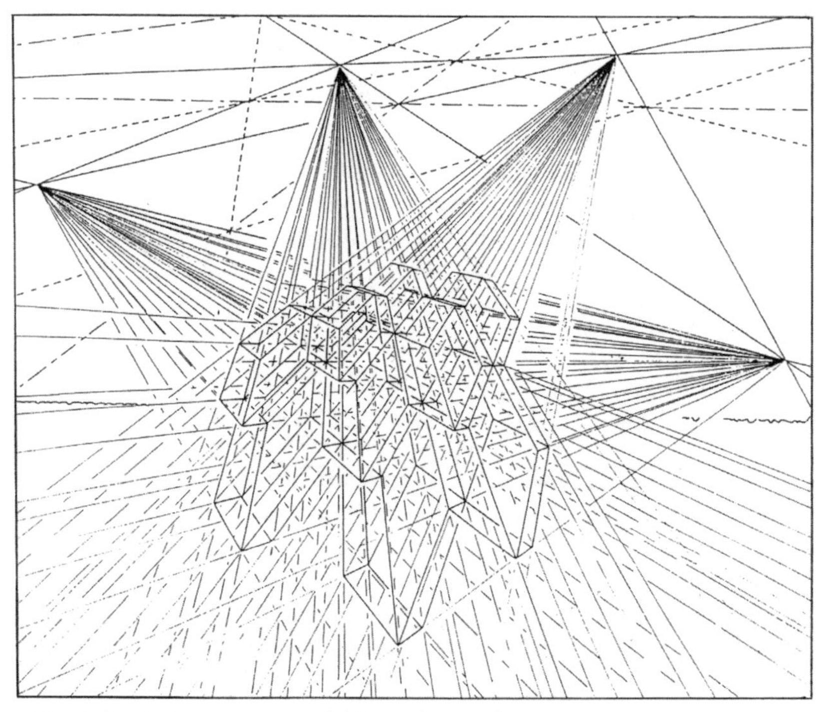

Fig. 2.3 Projective geometrical diagram for visualizing cosmic formative influences (from John Wilkes)

spiritual sources in the far cosmos is an awesome thought, yet crystals of different kinds are traditionally associated with constellations, even planets, and to this day are associated with various healing practices.[29]

How are cosmic forces transmitted?

Many readers will not see the necessity of associating cosmic forces based on the four ethers with measurable forms of energy reaching the earth. However, it could be argued that if such spiritual forces are to work creatively *on the physical plane*, they should at least be *conducted* by means that are detectable using existing instruments.

Detectable forms of energy travelling towards the earth cover the entire electromagnetic spectrum (Fig. 2.4). At one end there are radio waves (long waves of low frequency and low energy) which proceed through to infrared and the visible spectrum, then to ultraviolet, gamma and cosmic rays. The latter have frequencies falling within the range from cells to atoms and therefore have mutational potential. All these various waves or quanta of energy can barely be called physical at all but are candidates for *carrying* information. Sound waves, as Steiner enigmatically pronounced, 'do not themselves constitute tone, but are carriers of the tone which has its own separate existence'.[30] In modern telecommunications, the voice signal is superimposed on carrier waves. Just as electromagnetic waves are the basis of music, and as music carries meaning for the listener, it is by no means implausible to visualize such energies acting as carriers for forces which support life processes. As the most important form of energy received on earth is sunlight, there is a presumption that contained within the light from countless cosmic bodies are all the creative ethers (see below and Chapter 4).

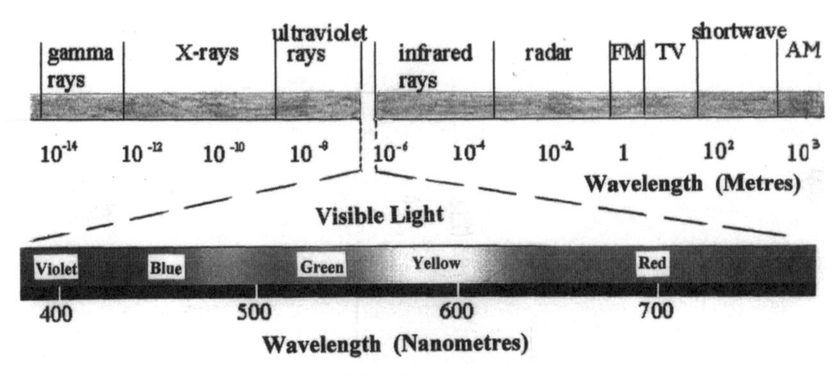

Fig. 2.4 The electromagnetic spectrum

Forms of human communication

Life for humans exists on the spiritual as well as the physical level, so how might spiritual communication occur? By definition it should not require physical means, and certainly in the realm of thought we can transport ourselves anywhere we wish in an instant! Seers or initiates, such as Pythagoras, the Holy Rishis or indeed Rudolf Steiner, together with other spiritual investigators to more limited degrees, could and can draw information from what are known as the akashic (cosmic spiritual) records. Among the abilities of Hopi seers and the shamans of other cultures is an awareness of the qualities of children before they are born. Lesser mortals can receive thoughts from those far away, from the deceased, and sometimes from angelic sources.[31] People, and particularly animals—including insects, amphibians and reptiles—have a sense of coming events. Consider the premonition of animals in the 2004 tsunami where comparatively few perished. Likewise the native inhabitants of Indian Ocean islands sensed unease and went to high ground. Pets are well known for sensing the return of their owners.[32] Certain healing practices, including radionics, demonstrate the use of mental or psychic energies and it is difficult to imagine that such activities require the services of the electromagnetic spectrum! Distant dowsing is in somewhat the same category.

Some have favoured the existence of a fine and unmeasurable web or 'field' of connections on which spiritual thought-forms and psychic phenomena are able to pass. By implication, a fine all-pervading 'ether' must be involved.[33] Such ideas approach reality but require spiritual insight to carry them further. For this writer, the existence of nitrogen in the earth's atmosphere is crucial, for Steiner has described this element as 'a cosmic substance here on earth, having will, knowledge, purpose and direct connection with groups of spiritual beings'. As it is our consciousness (astral body) which participates in the reception of tone and other information, the key chemical element *carrying* this spiritual principle is nitrogen. Nitrogen forms the bulk of the air we breathe, and nitrogen as protein and DNA is in every cell of our bodies. It is nitrogen which enables the mediation and decoding of cosmic spiritual forces. And nitrogen is also what facilitates access to a range of more worldly knowledge. We shall explore further the 'cosmic sensitivity' of nitrogen in Chapter 4.

The nature of our sun

The foregoing picture of our dependence on outer forces brings us inevitably to the sun, its character and crucial role at the heart of our solar system. To begin with, if we acknowledge that higher beings 'sacrifice' part of themselves for our benefit it is little use persisting with a view of the sun which lacks moral or spiritual credibility. Steiner advised in relation to stars that what we see as light is in reality 'the working of will and intelligence'.[34]

Meanwhile physicists understand the sun's heat and light as resulting from nuclear energy released from the so-called 'proton-proton chain' and also, to a lesser extent, from the 'C-N-O cycle'. They are increasingly sure about the processes that occur in stars, that different 'generations' of stars have formed since the so-called Big Bang (the supposed origin of the universe) that heavier elements can only form in stars of a certain mass, and that stars have different lifetimes according to their size.[35] However, the most important fact for our present purpose is their acknowledgement that *matter has originated from pure energy*, as is predicted by the Einstein equation $E = mc^2$.

If we consider the temperature pattern of the sun, at the surface of the photosphere it is around 5800° K.[36] While the core is hotter (Fig. 2.5), temperatures in the sun's plasma atmosphere, or corona—seen from earth at times of total solar eclipse—are around two million degrees. There is no satisfactory consensus for why this should be, and it defies the Second Law of Thermodynamics. Meanwhile, as the sun rotates, its outer layers do so

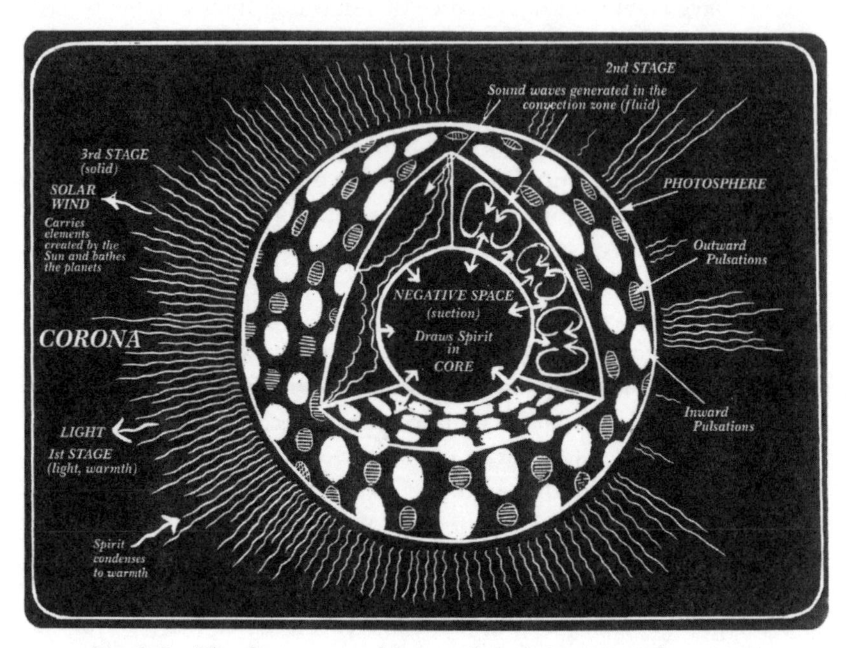

Fig. 2.5 The physiognomy of the sun (from Robert McCracken, 2001)

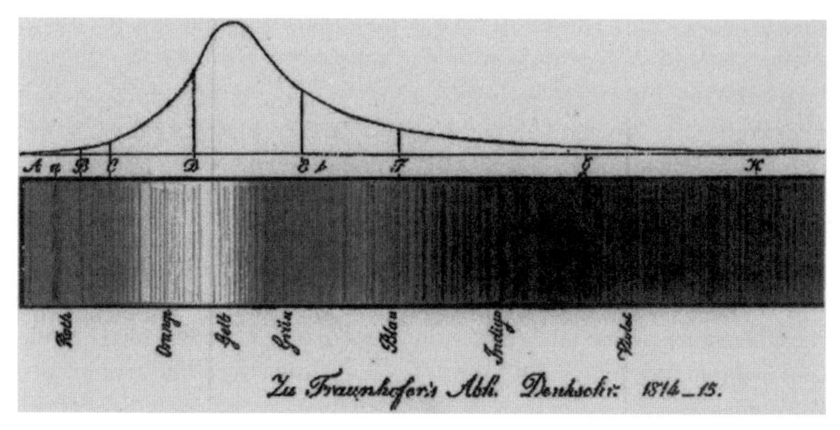

Fig. 2.6 Fraunhofer dark banding of the solar spectrum. Top curve shows light intensity. In this diagram from the original German publication, red is at the left and violet on the right

faster at the equator than at the poles so frictional energy is released. As a result, sound waves are recorded on earth which have bounced and reflected within the sun. This gives the impression of the sun being hollow,[37] an observation enhanced because the vast interior of the sun is almost entirely hydrogen (70%) and helium. Moreover, less than 1% of the sun's mass consists of elements heavier than helium yet these are concentrated not in the core, as would be the case on earth, but in the corona. We know from spectroscopy and the Fraunhofer dark banding (produced through elements absorbing light of different wavelengths) that the sun contains at least 68 of the chemical elements (Fig. 2.6).

The sun as gatherer and distributor of energy

Having established a few observable facts it is now possible to consider the views of Steiner, and those of others who have worked with his ideas. Amazingly for his time, Steiner rejected the 'ball of fire' notion and asserted that the sun's interior consisted of 'less than empty space'.[38] A helpful picture is given by Robert McCracken who develops further our opening thought. He says: 'According to spiritual-scientific law, the creation of warmth is a sacrifice of higher spiritual beings. When spiritual *inner* warmth becomes physically perceptible as *outer* warmth, a densification to air or gas occurs. Simultaneously a finer element is liberated—light. This is what is happening on the sun. Spiritual inner warmth streaming in from the cosmos condenses to outer warmth in the corona and liberates light at the sun's outer boundary.'[39]

But is this picture viable from an astrophysical standpoint? We can say that immense concentrations of energy (inner warmth) are drawn in towards the sun, manifesting as (outer) light and heat.[40] This process calls into existence subatomic particles, notably quarks, of which protons and neutrons are composed. In this way the smallest atoms, hydrogen and helium, can be formed, for it is these which most characterize the physics of the solar process. These substances are thus formed from a 'compression' of the surrounding stream of approaching energies. *In this way, matter is formed from pure energy*. But, according to McCracken, as spirit (or ether) acts in one direction, matter should move in the other. So the negative space or vacuum of the sun draws in cosmic ether forces while gross matter and manifested forms of energy migrate to the periphery—the corona.

Solar influences on the earth

The energy output of the sun is not constant in time, but undergoes shorter- and longer-term changes, notably the 11-year sunspot cycle. The sun thus emits a stream of high-energy material across the solar system which is directed to earth across the ecliptic plane. This is known as the solar wind (Fig. 2.7). The greater part of this stream is inimical to life and does not reach the earth's surface, but some enters via polar cusps in the magnetosphere, giving rise to the *aurora borealis* and *aurora australis*. The solar wind contains a number of life-related elements such as sulphur, calcium, silicon, iron, nitrogen, carbon, oxygen, phosphorus, potassium

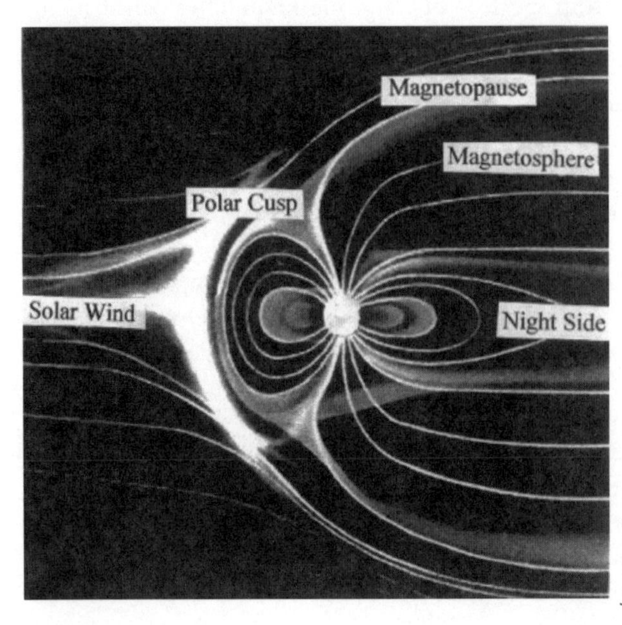

Fig. 2.7 The solar wind and earth's protective magnetic field

and manganese, which can therefore be considered to exist in very dilute form around the earth. Aside from direct radiation, global temperatures are affected by the strength of the solar wind (expressed as cosmic ray flux) for this appears to influence geomagnetic field strength and cloud development.

As the earth rotates, the day-side is subject to the strongest incoming energy. However, in etheric terms it is subject to the strongest attraction to a sun whose unusual character we have already drawn attention to. This forms the basis of the earth-breathing process introduced in Chapter 3 and its practical use in Chapter 5. When the moon or one of the inner planets cuts the earth's ecliptic plane, the stream of energies approaching the earth is disturbed and there are aberrations in the earth's electromagnetic field. At these times plant growth is adversely affected.[41] This is important evidence of cosmic influence on life processes.

For life on earth the visible sunlight carries life-supporting energies or ethers. We have mentioned these as originating from the zodiac, with the sun gathering, transforming and transmitting their energies. But from a geocentric perspective the sun's *quality* is modified by whichever zodiacal influence lies behind it.[42] This suggests that whenever alignments occur there is a channelling of life-formative or etheric forces. We shall revisit this idea in Chapter 6 when discussing the biodynamic calendar.

Many will see the sun as merely providing us with a lighting and heating system but in reality it is a vital intermediary between outer cosmic forces and life on earth. Moreover, the sun inspires a family of planets, which in turn influence life in particular ways.

The solar system

From Claudius Ptolemaeus of Alexandria in the second century AD until Tycho Brahe in the seventeenth, the earth was considered the *centre* of our universe. This is the actual experience of people on earth, for self-evidently everything moves around us! The fact that it was the sun which lay at the centre of a planetary system was first established by Nicolaus Copernicus. Rational observer he certainly was, but he nevertheless recognized each planet as ruling over a 'sphere' represented by its orbit around the sun, and that planets represented the activity of spiritual beings, of which the planetary body was their physical outer manifestation (Fig. 2.8). This is further illustration of a more universal mindset at the dawn of our present era.

Two particular 'motions' should be noted. While the planets form a family around the sun, in reality they are all *following* the sun, for the latter

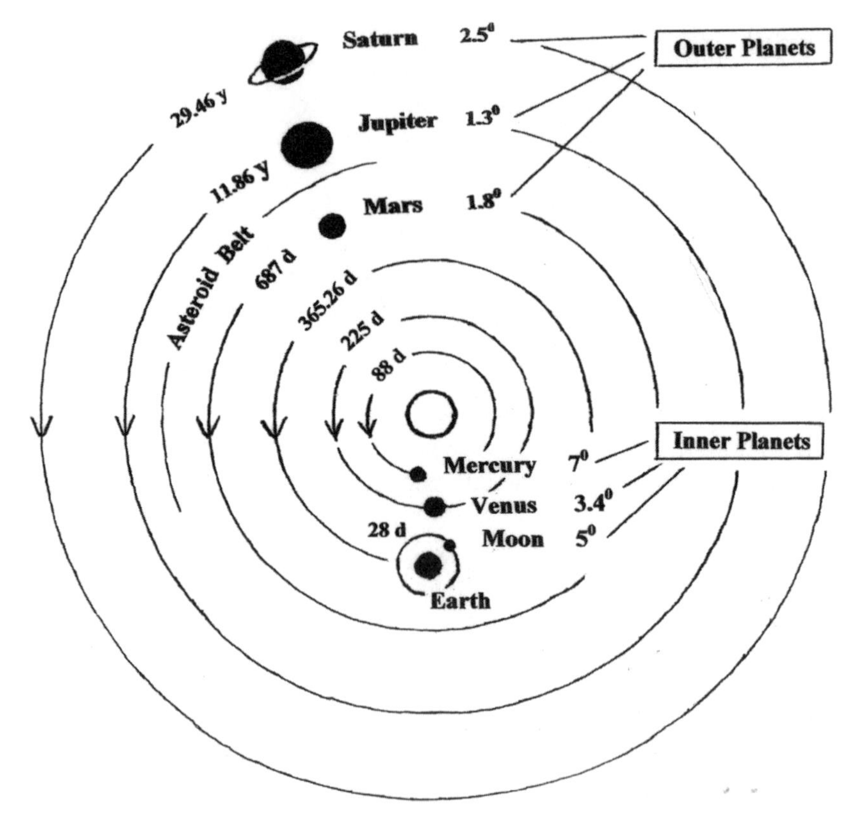

Fig. 2.8 Solar system showing orbital periods and inclination to the earth's ecliptic

is moving at immense speed towards the constellation of Hercules, the stars of which appear further apart as time has progressed. It is not surprising then, that planetary orbits, as Johannes Kepler discovered, are slightly elliptical.[43] Similarly, the moon's orbit is an ellipse, for the earth is in constant motion around the sun. However, it should be stated that ellipse theory is defective in so far as the forces required to maintain such motion cannot be explained.[44]

So in time this conception may be modified, for Willi Sucher, following Steiner's *Astronomy Course*, proposed that the earth and sun are involved in a figure-of-eight dance, known as a lemniscate.[45] This is one variant of a family of mathematical curves known as Cassini curves (Fig. 2.9). Lemniscate motion in some respects allows the conceptions of Copernicus and Ptolemy to be harmonized. Where a mutual attraction exists between bodies, both must move in relation to a neutral point,[46] an analogy being the somewhat circular motion of a man swinging a heavy

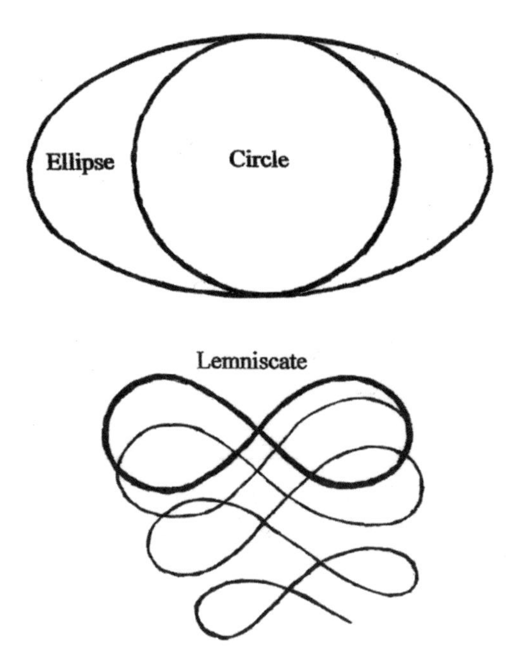

Fig. 2.9 Circle, together with examples of ellipse and lemniscate

weight. Speaking of knowledge within former mystery schools, Steiner said it was originally accepted that the earth 'forever moves to where the sun has been, reaching that point always a quarter of a year later'.[47] While this makes imagining the whole complex of planetary movements almost impossible, it is food for thought concerning the traditional seasonal festivals.

Because the planets have different orbital periods and different inclination to our ecliptic plane, no cosmic event can ever recur in exactly the same way. And despite the repeated cycle of our seasons, no two years can ever experience similar combinations of astronomical arrangements. Providing we can accept that supersensible forces influence us, this is most significant as a framework for evolution, for the future will always hold new possibilities.

Stars were referred to earlier as outer expressions of spiritual activity. In a similar way, all life on earth—the human body for example—is also the outer garment for the working of spirit. Likewise, we need to think—with Kepler—of the planets as outer manifestations of groups of spiritual beings with different tasks, part of whose current evolutionary mission is to work with the sun to support life on earth. Some readers will object that life as we know it is not possible on other celestial bodies! The idea to work with is that other beings may not require physical bodies and that it remains our task to experience the special challenges of a physical existence. In reality then, the sun sends forth its influences throughout the solar system. These are worked upon by different groups of beings before finally being received by the earth as formative impulses.

As if to underline their ordered tasks, the planets as far as Saturn are not just scattered at random distances from the sun but conform to a mathematical approximation known as Bode's law, where the average radii of their orbits increase in a constant manner.[48] Thus if we refer to the average distance of earth from the sun as 1 (the standard astronomical unit), then list all planets from Mercury to Saturn, including the asteroid belt, we obtain the distances 0.39, 0.72, 1, 1.52, 2.9, 5.2 and 9.55—the ratio between successive pairs yielding the series 0.54, 0.72, 0.66, 0.52, 0.55 and 0.54.

Could there be further significance in these figures? Earlier we referred to the music of the spheres. Music, of course, consists of tone and rhythm. The fact that there were seven original planets might imply the above ratios are analogous to tonal intervals.[49] We might also observe at this point that the periodic table of the elements shows repetition of properties according to a harmonic law and that in Steiner's classification of the ethers the 'chemical ether' is connected with 'tone'! When a violin bow is drawn across the edge of a metal plate on which a fine dust is placed, Chladni figures are produced which illustrate that sound has an ordering influence on matter. A final analogue with planetary motion is the orbital character of electrons based on energy levels.

Mystery connections between moon and earth

Exploring further the relationships between earth (microcosm) and the wider universe (macrocosm), a question periodically asked is whether it is possible in mathematics, geometry or architecture to perceive the working of the divine. The Great Pyramid has been considered an embodiment of the earthly microcosm. Its sides 'face the four points of the compass and mark the spot formerly regarded as the centre of the earth'. If we consider Fig. 2.10A, from the work of John Michell, the elevation of one of the pyramid's sides, with the apex angle as viewed at ground level, is first superimposed on its square base. A circle is then constructed with radius equal to the pyramid's height. Having done this, we find that the perimeters of square and circle are equal! The pyramid is thus 'a monument to the art of squaring the circle' and of 'promoting the union of cosmic and terrestrial forces by which the earth is made fertile'.[50]

The proportional measurements of moon and earth reveal a similar and remarkable correspondence. Let us consider Fig. 2.10B. If circles are drawn tangentially, with dimensions representing earth and moon, and squares are then drawn to encompass them, a 3-4-5 triangle is found to unite both squares.[51] Furthermore, as with the pyramid, the larger *circle* combining the two radii has a circumference equal to the perimeter of the larger *square*.

Let us now consider Fig. 2.10C from a numerological point of view. According to Michell, the number 3168 which relates to both circle

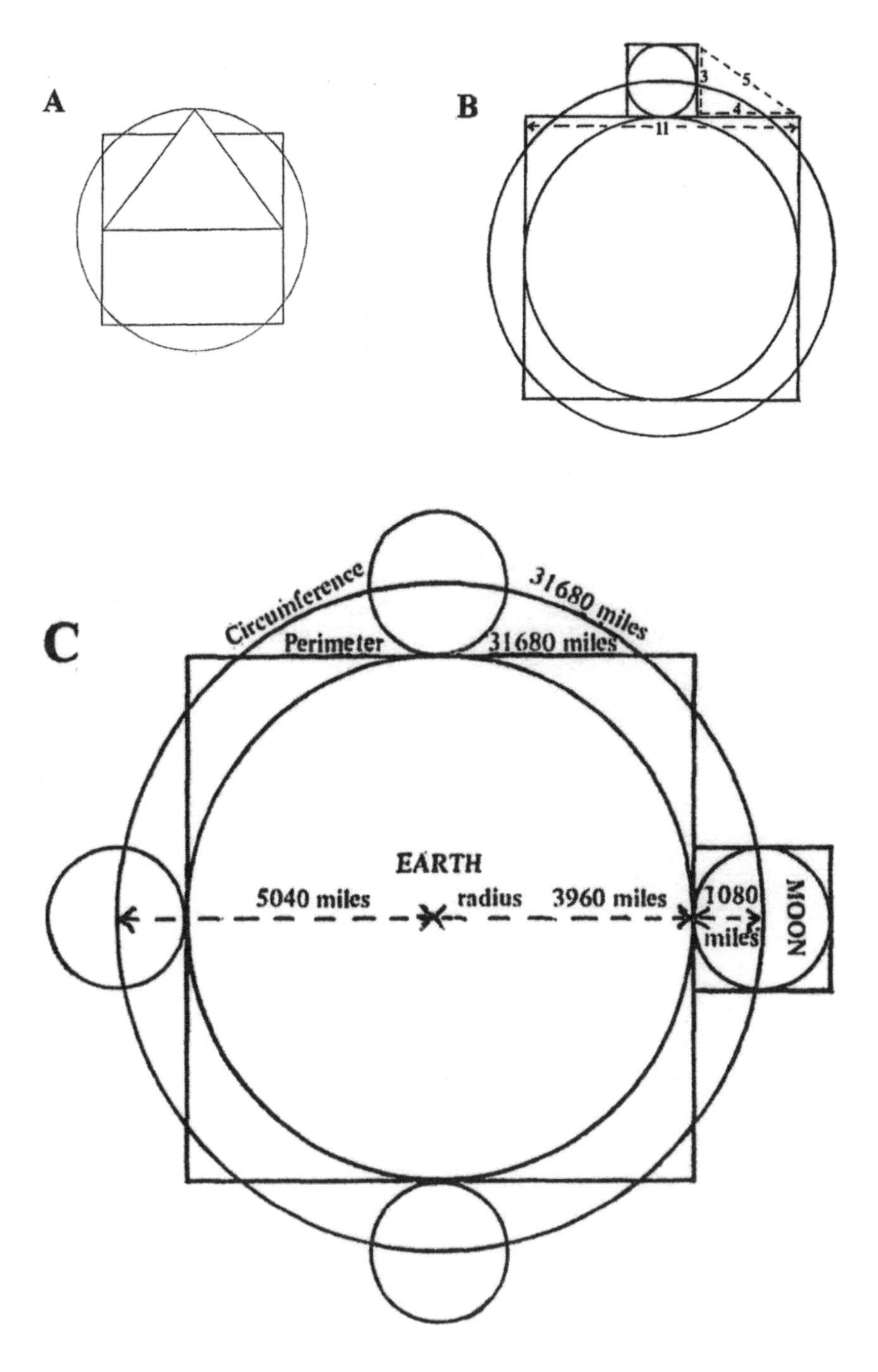

Fig. 2.10 Geometrical and numerological relationships between moon and earth (from John Michell, 1972)

and square is a Gnostic number. Such numbers arose through sounds having had creative power as well as meaning in primordial times. We have already referred to the significance of sound and will have more to say about it later. Sounds uttered were once able to be reduced to number. Thus in ancient languages, Greek and Hebrew for example, letters were assigned values (Gematria), these having cumulative meaning. Following this scheme, the value 3168 accords with the Greek equivalent of 'Lord Jesus Christ'. From Steiner's work we understand the Christ to be a cosmic being whose life, earthly death and resurrection was a defining point in his long and continuing involvement with *the whole of* humankind. In this respect it is of interest to note that the same numerals are found at Stonehenge, where the circumference of the massive Sarcen circle is 316.8 feet!

The combined radii of earth and moon (in miles) amount to 5040 (Plato's mystical number) while the diameter of the moon, at 2160 miles, is $\frac{1}{12}$ of the precessional period or Platonic Year, usually quoted as 25,920 years. When we then look at the average distance between earth and moon, it is found to be 60 earth radii (or 5, the number representing the human being, $\times$ 12).

Correspondences such as these inevitably arouse suspicion on grounds of selectiveness yet it has long been realized that British duodecimal units have their origin in an ancient cosmology. The reasons for drawing attention to these essentially mathematical relationships are that, while resulting from research into sacred geometry and numerology, they reinforce the notion that our current earth existence is overshadowed by cosmic wisdom.

Planetary rhythm and its effects on organisms

Another way of thinking about the solar system is that it imposes rhythm upon cosmic tone by the motions of each planet moving within its sphere. We should not think simply of movements around the sun, but subtle movements in relation to the earth. The latter all involve sequences of loops (see below). In this respect the earlier geocentric view of the solar system has closer connection with life processes.

People in ancient times had instinctive or intuitive awareness of nature, while their elites had wisdom imparted through mystery schools. From combinations of these sources a knowledge of the healing qualities of plants originated. Ayurvedic practice in ancient India recognized the influence of planets and their inherent rhythms on different organs of the body. It was the same in other native medical traditions. Such awareness enabled plants to be identified which had strong connections with a particular planet and which might provide cures if prepared in the correct way. Such knowledge only helps confirm the picture that the human being (microcosm) is created out of the cosmos (macrocosm). A straightforward example of planetary connection would be the yarrow, *Achillea millefolium*, formerly known as Venus's eyebrow. While the last vestiges of arcane knowledge are to be found in the *Herbal* of Nicholas Culpeper (1826), indigenous knowledge of the therapeutic qualities of plants is widely encapsulated in their Latin names, for example *Pulmonaria*, the lungwort.

While there is greater acceptance of *lunar* effects on plant growth,[52] ignorance or scepticism is widespread on the matter of *planetary* influences. Nevertheless, for the planets as far as Saturn, Kranich offered a framework for their influence on plant life.[53] He saw the upward-rising stem as responding to solar influence, and the downward root as reflecting the moon. The pattern of leaves as they diverge from the stem as well as patterns of flower petals were images of the movement of Venus and Mercury. Anther and pollen formation were considered a Mars influence while fruit formation was thought to be controlled by Jupiter, and seed formation by Saturn (see inner and outer planets below).

The patterns of petals, seeds and leaves have long intrigued botanists. Let us consider the symmetrical manner in which leaves or leaf stalks arise from the main stem—an example of phyllotaxis. Grohmann's example of the rose will be taken because its leaf symmetry is mirrored in the formation of petals, which rarely occurs in other families:[54]

> The leaves are arranged in a spiral around the stem (contracting towards the top) [Fig. 2.11]. The sixth leaf stands vertically above the first and to reach it one has to circle the stem twice. Seen from above, the shoot has five rows of leaves which correspond exactly to the angles of a pentagram.

In this arrangement we should note that the angle between successive leaves is $\frac{2}{5}$ of the stem circumference and careful examination will reveal that between each successive leaf one angle of the pentagram is always bypassed. Other plant families exhibit different numerical relationships: $\frac{1}{3}$, $\frac{3}{8}$, $\frac{5}{13}$, $\frac{8}{21}$, $\frac{13}{34}$. Such numbers are recognized as part of a sequence known as the Fibonacci series: 0, 1, 1, 2, 3, 5, 8, 13, 21, 34, 55 . . . a summation series where each successive integer is the sum of the preceding two. The ratio of successive numbers e.g. 21/13 tend eventually towards 1.618—the Golden Ratio[55] (see below). If we form two diagonals of the regular pentagram, this same value is achieved by intersection.

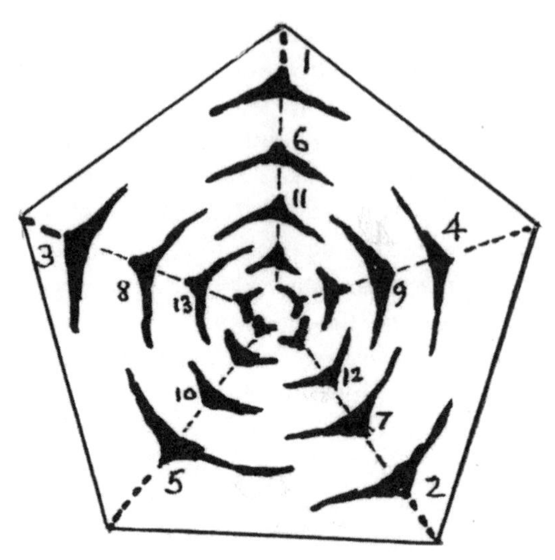

Fig. 2.11 Phyllotaxis of the rose (after Gerbert Grohmann, 1989)

Planets express this same numerical behaviour in the frequency of their close encounters with the earth. If we consider Venus geocentrically, we find that over a period of 8 years it accomplishes 5 loops towards the earth (Fig. 2.12).[56] Moreover, Venus rotates clockwise on its axis, *always* presenting the same face to the earth at the point of closest approach. This behaviour is indicative that the rose family (a representative of dicotyledons) is strongly influenced by Venus—a wonderful example of the effect of cosmic rhythm on life forms. Mercury accomplishes 3 loops every year (1/3 relationship) and its influence would appear to relate to monocotyledons such as grasses, alliums and the iris family. Further relationships have been proposed for other planets.

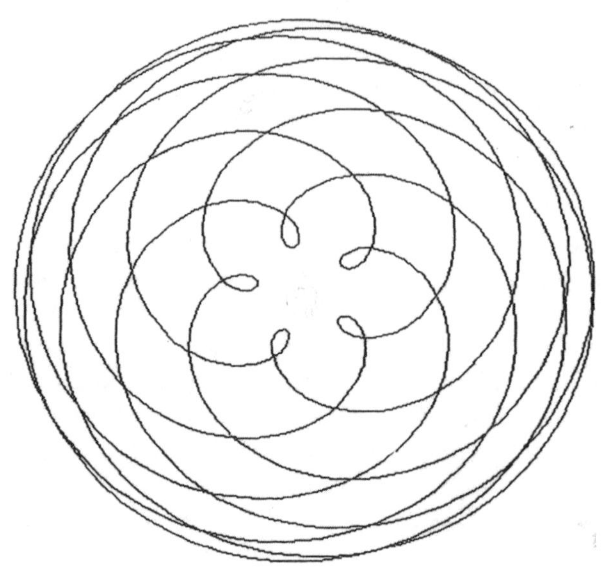

Fig. 2.12 The loops of Venus (from Brian Keats, 1999)

Phyllotaxis is also displayed by buds, cones, petals and seed heads and can sometimes be related to Fibonacci fractions. The meticulous research of Lawrence Edwards has shown that cones and buds respond to planetary movements. Buds of tree species expand and contract on a roughly fortnightly cycle as the moon aligns with particular planets. They are at their most expanded when earth, moon and planet are in alignment, and most contracted when the moon is at right angles to the line between earth and planet. But his greatest achievement, following George Adams, was in furthering the connection between projective geometry and living nature. Thus are organic shapes invariably bounded by characteristic linear forms called path curves.[57] For example, the shells of invertebrates exhibit a form of 'growth measure', a logarithmic spiral expanding from the centre and related through geometry to the Golden Ratio.

The Golden Ratio

Reference to this ratio should not be made without considering its derivation (Fig. 2.13A). We need to construct a 'root five' rectangle ABCD. To do this, the side of a square of unit value 1 is first bisected at E, and with radius EF (root 5/2) a semicircle is drawn. Two smaller rectangles, Golden Section rectangles, are produced, each with base 0.618 units. Each of these has a *reciprocal relationship* to the remaining rectangle of base 1 + 0.618. In other words, though smaller, they are similar in proportions.

Discovery of this ratio was a major achievement of Greek art and architecture, yet the same rules form the basis of organic 'growth measure' shown by the spiral phyllotaxis of sunflower seeds and by invertebrates such as *Nautilus* (Fig. 2.13B and C). Just how these living forms relate to the Golden Ratio and a rational system of numbers may be illustrated in Fig. 2.13D and E. In diagram D, starting with a 1.618 rectangle ABCD, we first create the perpendicular AF to the diagonal CB, intersecting at what we will call *the pole.* In doing this we have created three proportional triangles whose apices are the pole, and whose long sides are respectively AB, AC and CF. If we now make FE parallel to AC we have created *the reciprocal* of the larger rectangle and this whole process of reduction in geometrical proportion can continue to infinity, as indicated by the bolder line. In diagram E we have fitted a continuous curve to these same dimensions in order to create a spiral, characteristic of living forms.

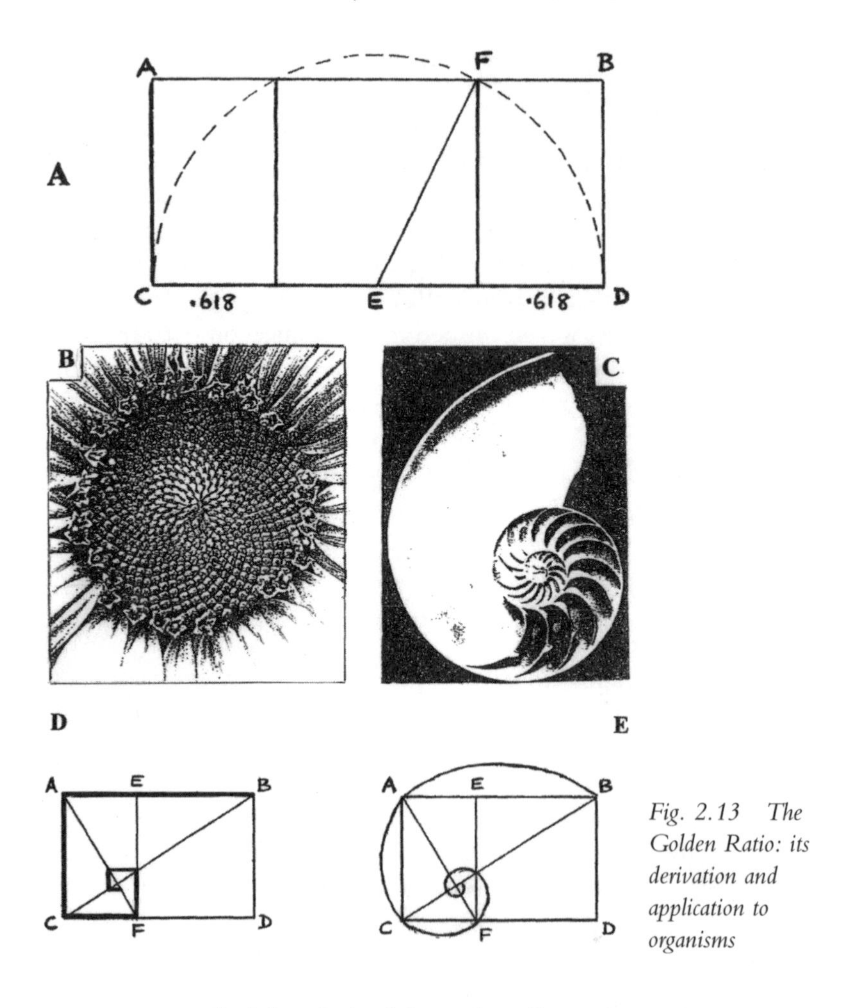

Fig. 2.13 The Golden Ratio: its derivation and application to organisms

Insights derived from the effects of sound

In the organic world, characterized as it is by curved surfaces, it seems reasonable to say that energies or forces of different origin may induce their own 'fields' just as iron filings reveal the field of a magnet. In this way, the formative effects of sound have been known ever since the effects of Chladni's violin bow in the late eighteenth century but I recently discovered an example of such effects by an author who had produced 'organic', mathematically related forms by vocal sounds. Different patterns were achieved according to certain variables, including the medium being used and the pitch of the sound. As Margaret Watts herself stated, 'we have only to examine these figures ... [Fig. 2.14] to find ourselves face to face with Nature in her almost limitless variety'.[58] To

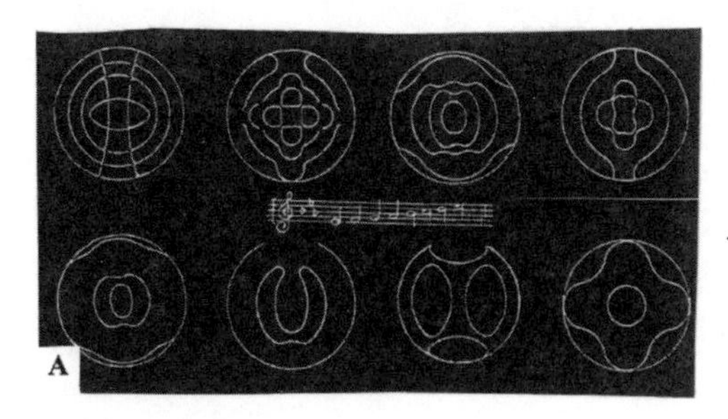

Fig. 2.14 Patterns produced by vocal sound (from Margaret Watts, 1891). A. Figures in sand according to pitch. B. Fern and tree forms using moist watercolours. C. Coral-like form. D. Linear form similar to bivalve. E. Floral forms (drawn because highly ephemeral). F. 'Cross-vibration' pattern (compare with Fig. 2.13B). G. Patterns in Lycopodium *spores*

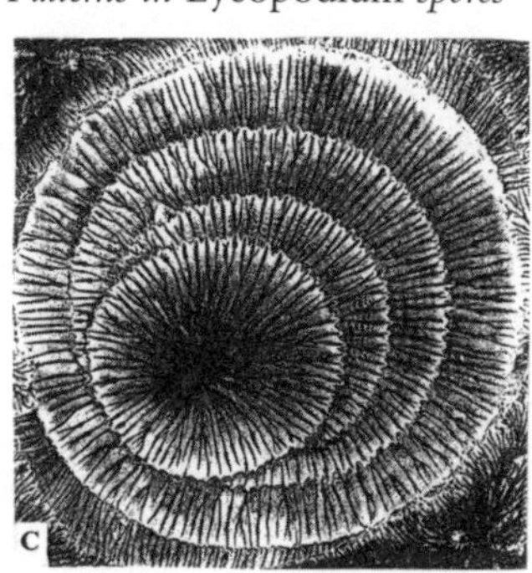

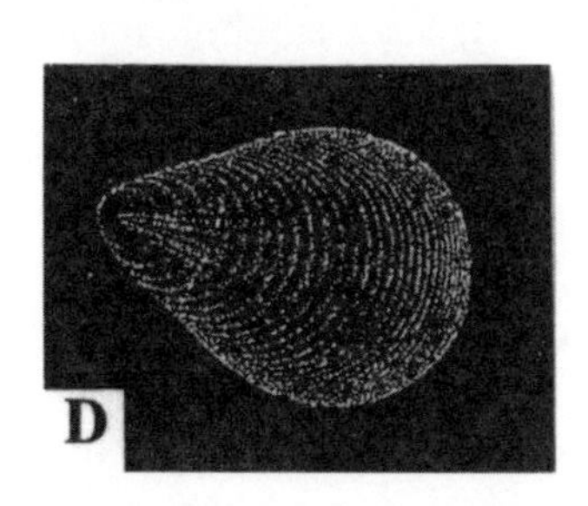

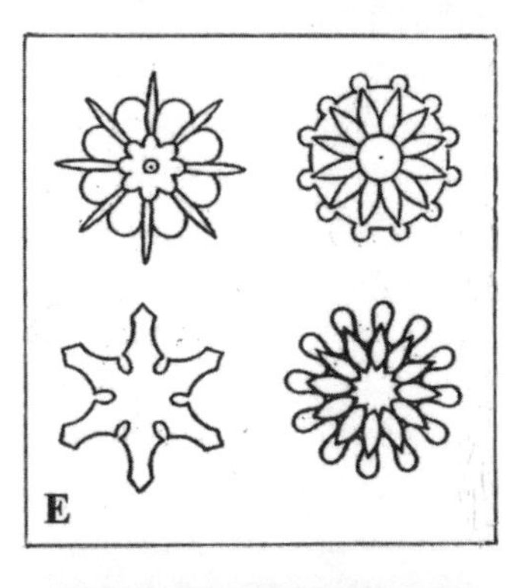

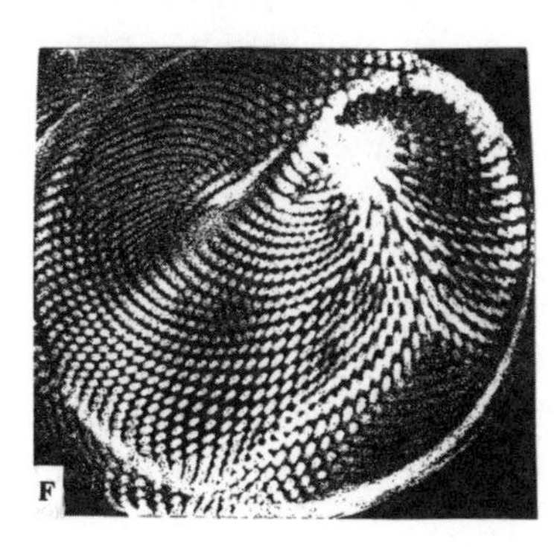

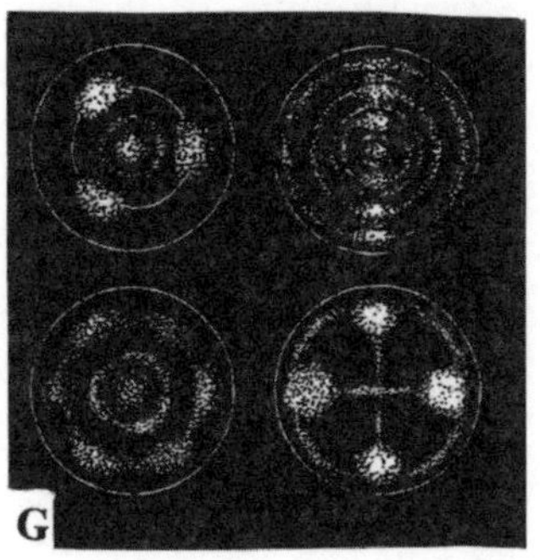

produce such effects in powders or liquid media clearly shows a relationship between vibrational energy and 'organic' form. It is a further step to see it as an analogue for the working of cosmic ether forces in the shaping of living forms.

Combining this evidence with that of the Golden Ratio and Fibonacci, we gain a glimpse of how cosmic ethers work according to mathematical laws in life processes just as they do in the formation of crystals in the mineral world.

The planetary days of the week

The recognition of planetary rhythm and its importance in human life probably found earliest expression around 4000 years ago in the designation of days of the week. It is almost certainly to the Babylonians and later the Assyrians, whose elite priesthood built observatories for studying the movements of the stars and planets, that we owe this seven-day system. Seven is a mystical number defining periods of time and it happens that the sun, moon and the then-known planets came to this number. Each day of the week bore either the name of the planet or the deity then recognized as ruling it. Aside from the Jews, all pre-Christian cultures were polytheistic and recognized planetary deities. Thus from northern European folklore, Tuesday to Friday are named after its eponymous deities: Tiwaz, Wotan, Thor and Freia.

What is remarkable, however, is not so much the list of sun, moon, Mars, Mercury, Jupiter, Venus and Saturn, but their order. Starting with the sun at the centre we progress to the moon which, in its association with the earth, is considered an inner planet, for part of its orbit does indeed pass *inside* that of the earth's around the sun. Furthermore, as we shall see below, it exerts typically 'inner planetary' influences. After this we have Mars, an outer planet (orbiting outside, or superior to, the earth's orbit around the sun); then Mercury, an inner planet; Jupiter, an outer planet; Venus, an inner planet; and finally Saturn, an outer planet (Fig. 2.8). The alternating sequence therefore embodies balance.

It may be objected that no mention is here made of the three furthest planets of our solar system. These are generally understood to have arrived later in its cosmology. The Tamil rishi Kakabhujandara more than two thousand years ago is reputed to have said that in the future the earth would experience the influence of three planets in addition to those then recognized. These were named in Sanskrit, Arakkan (Uranus), Samarassana (Neptune) and Kandakan (Pluto). Later astrologers were to observe that knowledge of electricity, magnetism and atomic energy were revealed to human beings by spiritual influences from these sources.[59]

Inner and outer planetary influences

In his agriculture lectures, Steiner emphasized the different roles of the *inner* (inferior or near) planets, moon, Venus and Mercury, and *outer* (superior, far or distant) planets, Mars, Jupiter and Saturn. We have already noted that the planets have a relation to life processes, this applying to animals as well as plants. The inner planets have a particular relation to digestion and metabolism, to growth and reproductive processes. In the case of plants, they support the role of the sun in photosynthesis, connecting predominantly with the vegetative processes of roots and leaves. They are therefore much connected with the process of incarnation—the accumulation of substance in the physical world, and therefore with production from an agricultural point of view. We should note the classical connection of Venus with fertility, Venus being linked to nutrition and its biological consequence, excretion.[60] Mercury was the messenger of the gods, and is linked to movement and healing. The moon is connected with fertility and all water processes. The inner planets therefore connect strongly with the elements of water and earth.

Inner planetary influences are drawn downwards into the earth's sphere by calcium and kindred substances.[61] Calcium is a major constituent of bony skeletons and is, of course, vitally present in mother's milk. Calcium is also present in nerve tissue, especially at synapses and the terminal points of cells. In plants, calcium ions mediate many different processes connected with cell division and cell signalling. These act structurally as a cement between adjacent cell walls and are involved in cell elongation, and so are particularly concentrated at meristems and root tips. Calcium is therefore of utmost importance in the process of growth, cell communication being a vital part of this. It is a veritable gatherer of the life or growth force. If we consider how widespread calcium carbonate is among living organisms we will note that in this substance calcium is combined with carbon, the form-bearer, and oxygen, the carrier of the life principle (see Chapter 4). Calcium is therefore implicated in linking life forces with physical substance. Calcium is also a representative of a group of substances we call salts, which the alchemists termed 'sal', and which draw life-giving water to themselves.[62]

What then of the outer planets? These could be said to manifest in a more subtle and perhaps qualitative rather than quantitative manner. We can again see them as supplementing the role of the sun, for example accounting for the colour and scents of flowers. Hence the spectral colours red, yellow and blue are connected respectively with influences from Mars, Jupiter and Saturn, a fact recognized by Hindus and Buddhists. There can be little doubt that such colour radiations work creatively elsewhere in the living world and colours are, of course, recognized as having therapeutic value.[63] Flower and seed formation represent the more cosmic aspect of plant growth so we should associate outer planetary forces predominantly with the elements of light and warmth. Outer planetary influences connect also with attributes which are not immediately manifest, such as general vitality. Whereas *substance* is emphasized in the contribution of the inner planets, here we deal to a greater extent with the *formative forces* underlying growth. These are the forces originating in the far cosmos and referred to elsewhere as archetypal plant forces. For the animal and human realms, outer planetary influences support sensory, instinctual and higher spiritual capacities.

Outer planetary influences find an affinity with the earth primarily through silica. They are first drawn into the earth before working upwards into the plant (see Chapter 4). Silica is a mysterious substance which Steiner describes as 'generalized outer perception' and as 'the universal sense within the earthly realm'.[64] Silica's relation to the physical world can often be described as 'on the periphery'. Silica is concentrated in the earth's crustal layers. In plants, while dissolved silica is present within cells, it occurs in grasses as peripheral opaline structures which add strength to the cellulose wall. A silica-rich membrane occurs outside the soft parts of a hen's egg and, similarly, outside the developing child in the mother's womb, and in skin, saliva, hair and on the surface of the eye. It can rightly be said that we 'look through silica to see the light'. In this way we build the picture of silica and calcium as polar opposites.

So how can we best summarize the roles of silicon and calcium? If we want to build a house we need an architect and a bricklayer. For growth and development of plants and animals we need formative forces and nutritional substance. We need these processes to occur in partnership. At the start of this chapter we made reference to the Greek concept of *zoë*—essentially an architectural force embodying the work that the sun carries out with the help of the outer planets and in connection with forces relating to silica. On the other hand the bricklayer involved with earthly life is the chemist and nutritionist connected with the sun's activity in relation to the inner planets and lime-related substances. This we have referred to as the physically manifested life or *bios*.

Achieving a balance in the working of inner and outer planetary forces is a major objective of biodynamics. It underlies the balanced healthy growth of plants mentioned in Chapter 1 and to be discussed further in Chapter 4. It is also the basis for production of food of the highest value for human beings (Chapter 10). The Lord's Prayer, which states 'Thy *will* be done on earth as it is in heaven' truly requires us to know what that will is, before we can set out to accomplish it in agriculture.[65] Recognizing the importance of these different cosmic forces is a crucial first step in achieving this.

3. The Living Earth and the Farm Organism

This chapter will place farms and gardens in the context of processes affecting the whole earth. The objective will be to deepen the farm organism concept as outlined in Chapter 1. It will inevitably be an exploration of much that is invisible to the human eye, but which is no less practical because of that. We will begin by considering the *earth* as an organism.

The human being and the earth

The idea of a living earth has a long history. In times past, human beings felt themselves *a part* of nature whereas now the majority have come to feel themselves *apart* from it.[1] So it is clear that a most significant step was reached when Descartes declared that the earth was without soul. Today, daring to think it does calls for a certain effort! In the past, people saw living organisms, together with the activity we associate with the atmosphere, oceans and earth's interior, as the expression of a planet with a soul[2]—aspects of the Greek notion of Gaia. In this soul life there may be little for most of us to see, but much that people can still feel. Moreover, just as an onion grows in soil and the air above, so the earth planet lives in its cosmic environment. For rationalist thought to arbitrarily separate life on the earth from its cosmic connections is a consequence of our present-day lack of spiritual consciousness. In fact, the human being and the solar system have evolved together through many cyclical epochs, and in this sense the earth has been prepared as a living organ for the evolution of humanity.

Formerly the earth was very much more 'alive', for in a previous epoch plant and mineral worlds were merged.[3] Though more highly differentiated than in the past and comparatively subdued in activity, one outward aspect of the earth's living character today is the seasons. These take different forms wherever you go but they are rhythmical, and for many of us, despite all-year-round availability of vegetables and fruits, the seasons still substantially govern our lives. So in different parts of the world leaf fall and fresh growth occurs annually, though the underlying factors may be different. This could be likened to a breathing process. Leaf development is accompanied by photosynthesis which absorbs carbon dioxide and releases oxygen, while respiration by countless organisms does the reverse—a complex of breathing processes, both seasonal and diurnal.

A key aspect of the earth organism for Rudolf Steiner was its spiritual breathing process, a process relating to the seasonal (and diurnal) position of the sun, and underlying all processes in nature. In this picture, the life of the earth and the elemental spirits which are an inseparable part of it are drawn out into the cosmos during the summer and withdrawn into the earth in winter.[4] Likewise, there is a daytime and night-time rhythm.[5] This principle applies according to the natural rhythms in different parts of the world. We shall encounter the same principle later in this chapter, and when discussing the biodynamic preparations in Chapter 5. This 'breathing', in an etheric or life-related sense, is a direct consequence of the nature of the sun as outlined in the previous chapter (Fig. 3.1).[6]

The human domain lies between the atmospheric realms of higher beings, mainly those of light, and those of sub-nature within the earth's interior,[7] and it is our task to keep some sort of balance. In ancient times, names were given to the respective spiritual forces or deities felt to be exercising their influence in these realms. In this respect, all matter, from the most ponderous to the most tenuous, is permeated by spiritual forces. Our 'atmosphere' forms an immense astral envelope,[8] partly of the earth but partly of a spiritual world too. Astronauts have reported strange and even life-changing experiences on their missions,[9] while adepts are known to reside in high mountain areas because of their more spiritual environment.

The biosphere, a relatively thin layer in which life occurs, is transitional between 'solid' earth and atmosphere. This we are seriously endangering, and ourselves in the process. While natural regions of the earth's land surface continue to be depleted and degraded, marine life is threatened as never before by over-fishing and changes in ocean chemistry, temperature and sea levels. Human beings have doubled the carbon dioxide (CO_2) concentration of the atmosphere in the last 200 years which has increased its so-called 'greenhouse effect'. In consequence, increased growth rates have occurred, moisture permitting, while some trees have responded by

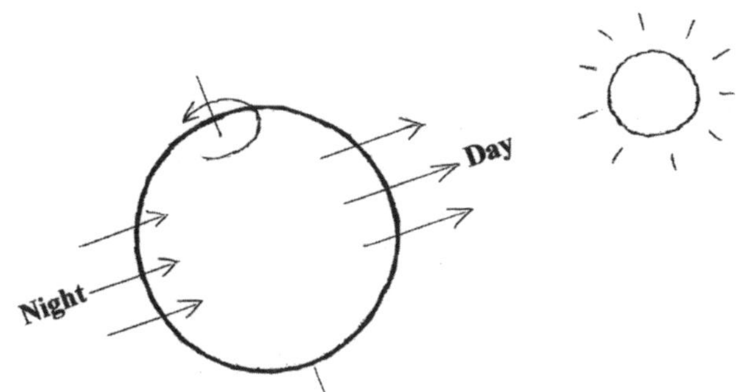

Fig. 3.1 Earth's breathing process and its relation to the sun

decreasing stomatal numbers. Much former increase in CO_2 was taken up by increased tree growth, but as the world is deforesting an area the size of Scotland each year this mechanism will no longer protect us. It has also become clear that nitrous oxide (N_2O) and methane (CH_4), released mainly from intensive agriculture systems, represent even more potent greenhouse agents.

These issues will appear to involve inanimate gases but, as mentioned above, spiritual beings are involved too, for *everything physical embodies a spiritual aspect.* So if we pause to consider the planet's increasingly hostile climates, we should take into consideration the impact of surface and atmospheric changes on the planet's elemental forces and breathing rhythms. But it is not simply our *actions* which affect these forces—our *thoughts* too are experienced by beings at higher levels than ourselves as well as those connected with nature processes. In this way, 'The essential causes of what happens on the earth do not lie *outside* the human being; they lie *within humankind*.'[10] Thus our thoughts and attitudes are capable of affecting all that lies ahead of us. This idea will appear more relevant as we take further steps to understand the farm organism.

In search of the farm organism

Having sketched a broader picture, let us now try to explore the idea of the 'farm organism'. The analogy with a living organism and the intrinsic connectedness of the various parts was introduced in Chapter 1. We can draw on the mixed farm as our model, as this was the basis of the rural economy up to the twentieth century, and it champions the notion, if not the reality, of self-sufficiency.

Aside from the vexed issue of energy needs in a modern farming context, the organic farm should be self-contained and be as close as possible to a closed system. This much is enshrined in texts covering the principles of organic agriculture,[11] and is an eminently defensible, resource-based explanation of the concept. Steiner himself said that a farm 'comes closest to its own essence when it can be conceived of as a kind of independent individuality or self-contained entity'.[12] The ideal organic or biodynamic farm should therefore maintain its own fertility and provide animal feed from its own production. An essential aim is for nutrients to be kept in balance. In this respect Steiner urged his audience to adjust animal numbers to the manure needs of their farms. A balance of sorts can, of course, be achieved by means other than animal manures—green fertility-building crops for example—so horticultural operations or stockless farms can theoretically meet this objective. But production units where major importation of fertility takes place should certainly be considered as unbalanced, even 'sick' organisms.

At the same time, farms are not natural entities; they have been created by human beings and, in terms of design and infrastructure, usually not by the present occupants. Many have worked well under previous economic and social conditions but their attributes are incongruous in the modern context. The ideals of organic farming are today often stretched to the limit by economic circumstances. Farming is tied in to macro-economic life rather than to local networks; and although we need to re-invent the latter, a living has to be made in the world as people find it. All of us learn to accept compromise and it is no different with organic farming on the ground. So the reality is that although one may have a vision, this all too often has to be compromised in the interests of making a living. Consequently fertility inputs or feed may have to be imported, just as they are in the case of seeds and other essentials. One may encounter circumstances where farms obtain inputs from within their local community, such as farmyard manure, horse manure and fresh or composted green waste. So, given appropriate checks, one should be prepared to accept this arrangement as within the spirit, if not the letter, of the organic farm concept.

However, if we are to accept the situation where organic farms import some part of their fertility or feed, it seems more important to understand the deeper reasons why the ideal of the closed system or mixed farm is of such importance. We start to do this by considering the contribution of farm animals.

The role of the farm animal

Farm animals are fundamentally dependent upon plants. The latter are primary organs through which a living earth receives influences from the cosmos—they are really sense organs for the living earth. Their green parts strive upwards to the light of the sun while their roots aim for the centre of the earth.[13] Animals and human beings are beneficiaries of such an extraordinary interaction.

What can we expect our animals to be doing? In Britain at least, for much of the year sheep and cattle are out on the land. As such they will be grazing the grassy sward and feeding on a succession of lesser species whose growth comes more at certain times than others. Cattle will be able to browse the hedgerow species—that will be a treat![14] And again, when food is scarce, hay or silage will be available. Hay from the previous summer contains a number of forage species and mixtures of ripened seeds, so the fact that it is conserved rather than fresh is more than compensated by its nutritious quality. Well-managed cattle grazing plays an important role in promoting plant species diversity through controlling coarse growth and spreading seed in the dung.

Animals should be grazed in rotation in order to limit parasite problems and to ensure that grass is evenly grazed and remains healthy. In this respect sheep are valuable for cleaning-up pastures with much dead grass mat. Pigs are different, for more of their food, such as whey, is brought to them, but they can forage outside if space is available. Along with pigs, hens can be employed as gardeners to clear rough land. Otherwise hens, and also horses, need to be moved frequently in order not to degrade pastureland. While soil compaction can occur with cattle in the wet, pigs will relish wet conditions and are no respecters of soil structure! But most important is that animal dung and urine will be available to the soil, meaning the animals participate in a nutrient cycle, stimulating soil organisms by their presence. Let us look at this more closely.

As indicated, plants assimilate substances in their leaves through interaction between earthly substance—soil nutrients, water and atmospheric carbon—and cosmic energies, as sunlight. Everywhere on the earth's surface is unique. The geological formations and soils, the shape of the land, the character of the climate, the past history of plant life and land use—all this is special to each location. When a farm animal grazes, it consumes something incomparable which belongs to the animal's experience of life on that farm. For this reason, the dung it produces belongs to a farm in a way that the dung from feed that has been imported does not. And by extension of this argument, to import dung from elsewhere is to bring in organic substance which has no connection to the place where it is being used.[15] In terms of material input of nutrient one cannot see why such a difference is of importance but from a holistic standpoint, in terms of soil fertility and animal health, the picture is different. As disease periodically threatens the disappearance of animals from British farms we could do well to think a little more deeply about such matters.

The inner attributes of animal dung

A familiar question for organic producers and those considering organics concerns how essential it is to use animal dung in the compost for maintaining soil fertility. This is of most concern in horticultural operations where organic matter is overwhelmingly of plant origin. From a nutrient point of view, the same fertility input can be provided by decomposed plant matter. But animal dung is not simply what outer appearances might suggest.

The plant incorporates substances into its tissues, which participate in living processes. When the animal feeds on the plant, it feeds not on substance alone but on life energies or forces contained in the food which its digestive system subsequently releases. Meanwhile, the animal is a sentient creature with consciousness. The substances circulating within it are thus 'raised' to a level above what they were in the plant—they are exposed to further cosmic (planetary-mediated) influences. *Through this connection, animal dung and urine confer on compost and the soil a greater capacity, or sensitivity, for the working of cosmic forces in agriculture.*[16] The consciousness of the animal (astrality) thus becomes a 'consciousness' in the soil for all the cosmic forces that penetrate the earth. So even with the use of modest amounts of manure the value of compost can be enhanced (see below). If we stop to consider chemical fertilizers, which have no relation to life processes, their use provides plants with only a limited part of the stimulus required in order to produce crops containing real nourishment (Chapters 4 and 10).

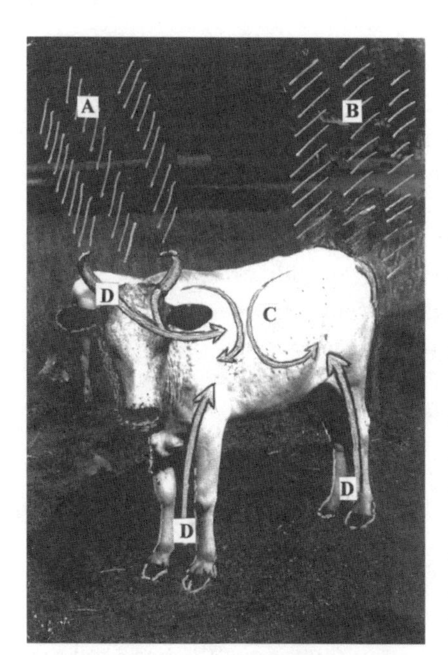

Fig. 3.2 Astral influences experienced by the animal and their intensification by horns and hooves. A. Outer planetary influences received. B. Inner planetary influences. C. Circulation of radiations within the body. D. Reflection back of dissipating energies

Cows and horns

We need to explore further this relationship of animals to cosmic energies. Influences mediated by the inner and outer planets penetrate their organism. In broad terms, the hind part of the animal is the conduit for energies from the near planets while the head region receives those from the far planets (Fig. 3.2). The chakra points recognized in human beings are similarly places associated with currents entering the body. Such currents can be considered to be in a constant state of inflow and outflow. Coordinated by the body's endocrine system, these currents—vital to the digestive process (Chapter 10)—can be seen as creating a resonance with the genetic material of every cell in the body.[17]

Differences occur between

females and males, and we need to address this in order to understand why, in biodynamics, we use cow manure and cow horns when making certain preparations (Chapter 5). Let us mention certain obvious points. Despite many people's reluctance, we can slowly and calmly walk through a herd of horned cows and there is little response—the munching goes on, and dreaminess in the eyes can be detected. The cow is entirely given up to the digestive process, an act of inner contemplation. By contrast, the bull not only scans the horizons for competitors but instantly knows when cows are ready to be served. The attention of the male is directed outwards in eyes and nose, while the forces of the female are directed inwardly. This is a rule with general application, for in the case of the female outer influences are retained more strongly in the organism while in the male they tend to dissipate more completely. The characteristics of hair growth may also be related to this. In the present context, the relative intensification of these forces in the body means that the dung from *any* female animal has greater value than that of the male.[18] But in the case of bovines we also need to consider the role of horns and hooves.

Why did most breeds of cattle have horns? We might consider that they were for armament and protection, and I am sure this is what farmers who have tangled with horns still feel! In the first place a bovine horn has a complex internal structure—not like the antler of a stag. It consists of an outer shell of highly compressed protein substance which, like our nails, is the termination of skin-formation.[19] The core of the cow's horn, which incorporates sinuses for the passage of air and blood vessels, is a bony extension of the skull. The great variety of horn types from cattle breeds around the world is testimony to the formative forces of these animals working in tune with local conditions. We can also say that the morphology and structure of male and female horns is different; bulls' horns are frequently straighter than the spiral form of the cow, they usually have a thinner shell and, more significantly, they do not exhibit such a complex internal structure.

We should also realize that horns only begin to grow as the period of suckling draws to a close and the calf begins to eat hay and grass—material with high roughage content. This is the time when the rumen begins to function. We have here a picture of the real function of the horn and can begin to understand why it should be necessary for the cow's breath and blood circulation to extend right through to the organism's extremities. However, horns and also hooves reflect back energies which would otherwise dissipate, thus intensifying the digestive forces of the animal (Fig 3.2). In the case of the cow, this intensification serves processes through which the calf and its milk are formed—it also means that cow manure is a powerful source of radiant forces.[20]

Why do we find a golden sun disc placed between the horns of cattle in Egyptian art? Is it any wonder that in India from ancient times the cow has been venerated as 'the goddess of fertility'? Serious disorders such as foot and mouth disease have increased in modern times as we have redesigned our herds to suit modern practices, while creating levels of stress in them unimaginable in former times.[21]

These are the issues which underlie the importance accorded to the cow in biodynamics and which underline the role of bovine animals, in particular, in creating and maintaining a healthy farm organism.

The use of animals and animal organs in biodynamics

The requirement for animal organs in order to make the biodynamic preparations (Chapter 5) has generated various concerns in recent years. Not only has the availability of suitable animal parts been restricted by EC and Defra regulation following the disastrous BSE episode, but the use of such organs presents an obstacle for many people.[22]

Meanwhile, some would question whether raising animals for food should be condoned, regardless of the agricultural system. No useful purpose is served in mounting a defence of biodynamics, for wider issues are involved.[23] Animals have served the needs of humanity from the beginning, while domesticated animals have owed their existence to farming activities the world over. This can be viewed as a sacrifice which the animal kingdom has offered and continues to offer. In a future epoch, according to Rudolf Steiner, it will be our task to further their evolution. Although a long-term view, this nevertheless represents a serious philosophical position.

It should be stressed that in organic systems particular attention is focused on animal welfare, including feed quality. The contribution of farm animals to soil fertility is self-evident, so the greater part of organic farming would be unable to function without animals as part of the farm system, or of someone else's. In the vast majority of circumstances, organic farming—indeed a huge swathe of agriculture—would not be viable without an economic value attached to farm animals.

A more fundamental argument is that without the animal element, ecosystems are incomplete, and without the contribution of animal manures the same sensitivity cannot be imparted to the soil. If the future of farming is to be more sustainable then it must embrace a *more* rather than less holistic concept of the ecosystem. As such, animals should be multipurpose, playing their part in farm production, fertility maintenance and, in certain instances, the control of pests and predators. An energy or soil-conscious world might also envisage their return to use in farm work and transportation. Aside from questions of dietary needs or preferences, it is sadly the case that extreme vegetarian or animal rights arguments end up unravelling the whole idea of organic farming as currently practised. The consequence is that we are left with an emaciated ecological option with which to confront less desirable farming alternatives.

Arable and horticultural rotations

Crop rotation was mentioned in Chapter 1. It is a traditional way of conserving soil fertility and avoiding pest problems, normally beginning with heavy feeder plants followed by those which are less demanding. A fertility-building (green manure) crop is then introduced at the end of the cycle, or at the end of each season in order to lift nutrients, prevent weed colonization and generate organic biomass.

In practice, rotations are a compromise between the needs of the farm for production of key crops and the need to look after the soil. The latter depends on maintaining a suitable physical structure as much as balancing the consumption of minerals. For a farm, it is most easily achieved by putting the land under a grass-clover ley. Nutrients can be returned to the soil by various means but the physical condition of the soil can only be maintained by a careful cropping policy. The ley allows an opportunity for the soil to build its organic matter content while usually benefiting from the presence of animals as described above. This contributes to a healthy farm organism for it helps maintain a diversity of soil organisms and will also improve the working of cosmic forces, to be explained in Chapter 4.

In principle, garden rotations are no different but many people's rotation is compromised by available planting area and the desire to produce a few preferred crops. We find also that in the garden much larger amounts of compost are deployed. With covered cropping, too, the concept of the rotation is challenged. Without saying that the soil here is artificial, its fertility is mostly maintained by barrow-loads of compost rather than from minerals slowly weathering from the soil underneath! One might say that there is less need for rotation under these circumstances but in view of the fact that these are intensive growing environments, diseases must be carefully watched. There is always a risk, particularly with mild winters, that disease and pest problems are carried over from one season to the next. So rotation is a principle lost at one's peril.

Looking at rotation another way, we harvest crops which emphasize particular plant attributes—mostly the root, leaf or fruiting organs. In this way we can adopt a rotation which recognizes the different elementary forces coming from the zodiac (Chapter 6). Such a scheme forms an ideal basis for a four-year rotation. Here, one feels that relationships between earth and cosmos are kept in balance. However, owing to disease problems running in families, as with the brassicas, a sequence based on root, leaf and fruit suggests that the actual succession of crops should be chosen from *different plant families*.[24]

The hidden qualities of compost

Compost operations should also be considered as organs within the body of a farm or garden, for key organisms participate in this process. Aside from farmyard manure stacked from the byre we could consider some typical compostable materials widely available at different seasons. These include recent crop residues such as straw, together with weeds, grass clippings and kitchen waste. The gardener may occasionally have access to some animal manure; in the absence of cow manure, pig, goat, horse or chicken manure can be used, the latter often available in dried form. The composting process will depend on the quality of these raw materials, notably their carbon to nitrogen ratio (see below), and how the heap has been formed. Compost will be bacterial if much fresh material has been used, and fungal if the material is all dead, as in the case of a pile of autumn leaves. A pile on which bits of rubbish are periodically thrown—typically the household bin—will develop more slowly as a worm compost.

We must understand what dealing with living matter really involves. What we call 'life' does not simply vanish when materials are placed in a

Carbon–nitrogen ratios of some common materials

Material	Ratio	Material	Ratio
Urine	0.8	Non-legume hay	30–35
Manure heap seepage	1	Kitchen waste	20–40
Dried blood	8	Water plants/weed	20–25
Chicken dung	8	Potato haulms	25
Sewage sludge	8	Fallen leaves	45
Bone meal	8	Shrub trimmings	50–60
Farmyard manure	11–18	Maize/millet stalks	70
Brewers grains	15	Cereal straw	80–100
Legume hay/straw	15	Rotted coir dust	110
Fresh grass	15–20	Coarse coconut fibre	300
Raw compost	15	Rotted sawdust	200
Mature compost	20	Wood shavings/paper	500

compost heap. When the original forms decompose, it is not just the substances which are released but also the forces which held those forms together. It is the tendency for this life force to dissipate which is the stimulus for proliferation of micro-organisms. A compost heap formed all at one time thus begins to heat up. Heating to around 60°C has the advantage that pathogenic organisms are killed, together with a proportion of weed seeds. In biodynamics we add a number of preparations (see Chapter 5) that control the breakdown process and help retain life forces within the compost.

These forces are valuable for subsequent plant growth and can be held more effectively by adding lime, dolomite or wood ashes to the layering, as well as by creating a domed shape for the heap or windrow.[25] We may recall from Chapter 2 that lime has an affinity for life forces, and for this reason it acts to dampen the tendency towards rampant decomposition. In addition, bacterial activity generates acidity through the production of nitrate from ammonium-nitrogen. Use of liming materials prevents compost becoming too acid and improves the subsequent activity of earthworms. As it is also beneficial to maintaining good soil structure, adding it to compost saves us adding it directly to the land, which is always a good idea.

Earthworms appear in the later stages of compost which has passed through a thermal process, helping to create structure and consistency within the material. On the other hand they are the key players in a cool composting process, often rather poorly orchestrated by the gardener! In soil they are the great distributors, they consume organic matter and its included bacteria and excrete it in a different place, some producing casts on the surface. In this way they help spread out and redistribute concentrations of life activity within the soil.[26] Without such work, plant roots would not grow healthily. Earthworms also secrete calcium from a gland in their gut which helps complete a picture of finely dispersed lime holding together life energies within the soil (Fig. 3.3).

So when we add compost to the soil, besides the benefits enumerated in Chapter 1, we distribute energies which radiate, and which support

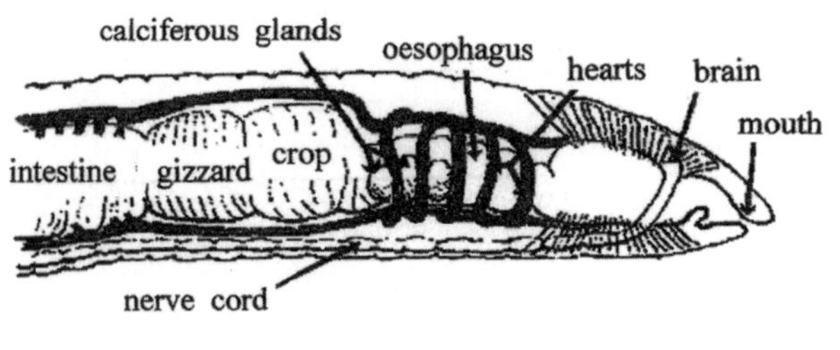

Fig. 3.3 The anatomy of the earthworm—head section

plant growth. The further incorporation of animal wastes, together with the compost preparations, means that when biodynamic compost is added to the soil the latter becomes more sensitive to cosmic forces and plants have greater capacity to satisfy their nutrient needs (Chapters 4 and 5).

There is no doubt that the value of compost will be maintained if it is not allowed to become saturated by rain and therefore subject to loss of nitrogen (as nitrate) through leaching, or through anaerobism (as nitrogen gas). Equally, the characteristics of organic carbon change irreversibly when allowed to dry out, so it gradually becomes of less value as a soil conditioning material. For these reasons, if we are to treat the compost operation as an organ of our farms or gardens, then the heaps must be covered to allow the material to mature without deterioration.

Misconceptions about micro-organisms

There continues to be interest in the subject of compost acceleration, not least by those supplying the cultured and processed material.[27] Many harbour the vain hope that by using these materials compost will be finished quicker, with less work involved. In the majority of circumstances where farm or garden waste is involved, the importation of organisms is unnecessary.[28] The truth is that the rate-determining factors for compost-making remain temperature, moisture, aeration and C:N ratio. In addition, compost is produced more rapidly if the fragments are initially chopped and the materials mixed. Unless conditions are strictly optimized for introduced organisms, there can be no guarantee they are the ones proliferating in the days subsequent to their addition! The fact is that bacteria and fungi, the simplest cellular forms of life, change their morphology according to ambient conditions, so composting can always begin from organisms naturally present on the surface of organic matter.

Effect of mycorrhizae on the nutrient content of maize shoots. (Average μg per plant, after Lambert, 1979)

	Without Mycorrhizae	With Mycorrhizae
P	750	1340
K	6000	9700
Ca	1200	1600
Mg	430	630
Zn	28	95
Cu	7	14
Mn	72	101
Fe	80	147

Furthermore, in most operations, speed is much less important than having *enough compost* when it is needed. And speed is less important than the creation of optimal compost. If a stimulus to compost-making is required, 'Mäusdorf' or 'cow pat pit' has been found to promote a satisfactory process (Chapter 5).[29]

Similar misconceptions apply when we consider the inoculation of plants, or rather their roots, with fungal mycorrhizae. There is no doubt that plants struggle to achieve good growth if denied root-symbiotic fungi (see above). Not only do these extend root systems by hundreds of times, they provide the plant with a specialized method of obtaining required nutrients and confer other advantages such as disease and drought-resistance. In consequence, endo-mycorrhizal innocula are widely available. While there may be pressing commercial reasons for taking this course, we have to wonder how people managed generations ago! We should realize that plant roots release exudates which represent 'carbohydrate signals', and that in a soil or planting pit which contains good compost a range of fungal types will naturally tune in or adapt to the particular plant. In short, the way we manage our soils affects how well nature's workforce can respond to our needs. It is a sign of mechanistic thinking to imagine that by inserting an organism all will be well.

Use of organic fertilizers on the land

A question often asked is: How much compost or farmyard manure do I need to add to the land? Frustrating as it may sound there is no single or simple answer to this! Aside from the matter of inherent soil fertility or the choice of crops to be grown, there is a fundamental issue in organic farming, still more explicit in biodynamics, that organic fertilizer is as much a *catalyst* to natural processes as it is a direct supplier of nutrients. This is the reason why the Demeter standard for nitrogen inputs is lower than that required under the European regulation. An attempt is made in Chapter 4 to resolve this enigma, but for the moment it will suffice to say that the organic nitrogen content of soil is crucial to harnessing cosmic energies. Another approach is to say that when fertilizer needs are calculated for crops under conventional farming there is likely to be a closer connection between these and crop performance than there would be for inputs and yields under organic systems. When we use organic fertilizer it may nonetheless be useful to know what levels of nutrient are being added. Samples of available data are presented below.

Much interest has centred on the cow. In the course of one year, a 450 kg animal is likely to produce 10–12 tonnes of solid manure and

Typical composition of manures and compost

	Dry matter %	Median % fresh weight		
		N	P	K
Farmyard manure	23	0.6	0.15	0.6
Poultry manure	29	1.7	0.6	0.6
Cattle slurry	4	0.3	0.04	0.25
Pig slurry	4	0.4	0.09	0.17
Sewage sludge	5	0.25	0.16	0.01
Compost	28	0.9	0.22	0.33

Fertilizer value of one tonne of compost

Composition	**Fertilizer equivalence**
30 kg Nitrogen	150 kg fertilizer at 20% N content
20 kg Phosphate	110 kg superphosphate
25 kg Potassium	65 kg potash at 40% K content
375 kg Calcium	750 kg agricultural lime

Micronutrients in farmyard manure

	Mn	Zn	Cu	Mo	B	Co
Kg elements in 45 tonnes FYM per ha	3.36	1.12	0.56	0.02	0.23	0.01
Kg/ha requirement of 4 successive crops	2.5	1.8	0.3	0.02	n.d.	n.d.
g/tonne of each element in FYM	18–137	10–53	2–10	0.2–1.05	1–1.3	0.05–1.2
Mean uptake by Plants g/ha/yr	500	200	80	10	180	1

5–6 m^3 of urine which, with bedding straw, comprises farmyard manure. In other estimates, it is said that cattle give 10–15 tonnes of solids and liquids when kept inside for 6 months, while outside they produce around half this amount. One cow in one year, depending on nutritional regime, produces around 50 kg N, 15 kg P and 50 kg K *via* farmyard manure, while urine additionally accounts for 40 kg N, 2 kg P and 60 kg K. Just because organic farmers do not go around with sacks printed with percentage nutrient content, it should not be assumed they are grossly under-fertilizing. This brings us finally to some recommended rates of application for organic and related materials (see below).

Recommended application rates for organic fertilizers

Farmyard manures (IFOAM)

Cattle manure	20 tonnes/acre (strawed)
Cattle slurry	12–16,000 litres/acre (undiluted/diluted)
Pig slurry	8–10,000 litres/acre (undiluted/diluted)
Poultry manure	2–4 tonnes/acre (deep litter/free range)

Compost amounts in relation to soil type

CROP TYPE	LIGHT SOIL	HEAVY SOIL
Heavy feeder	20 t/acre (4.5 kg/m^2)	15 t/acre (3.5 kg/m^2)
Moderate feeder	15 t/acre (3.5 kg/m^2)	10 t/acre (2.2 kg/m^2)
Light feeder	1–2 t/acre (sprinkling)	0.5 t/acre (sprinkling)

Intensive greenhouse/polytunnel system: 30–40 tons/acre (6–10 kg/m^2). For general home gardening, one household bucket per square metre is a good guide.

Other materials

	TO COMPOST	DIRECT TO LAND
Calcified seaweed	3 kg/m^3	50 g/m^2
Bentonite	5 kg/m^3	100 g/m^2
Basalt rock dust	10–15 kg/m^3	100 g/m^2

Liquid organic fertilizers and compost teas

Organic and biodynamic farmers and growers use a range of home-made liquid organic fertilizers primarily for foliar application. Often referred to as liquid manures, these consist of dissolved nutrients and organic colloids in suspension. The production of liquid fertilizers will often depend more on fermentation than on aerobic breakdown of solids as in a compost heap. Just as with composting, many different local resources can be used to create a liquid manure, including animal dung, especially from cows or chickens, green leaves such as nettles or comfrey (leguminous leaves are widely used in the tropics), pond weeds and algae, fish remains and seaweed.[30]

Liquid fertilizers deserve to be widely used because they offer more immediate response to nutrient deficiencies than via the soil. They are also another way in which the resources of the farm or its immediate neighbourhood can be harnessed with minimum cost or energy use. There is evidence that these liquids, especially when enhanced with certain herbs, can strengthen plants against pests and disease. Compost teas and liquid

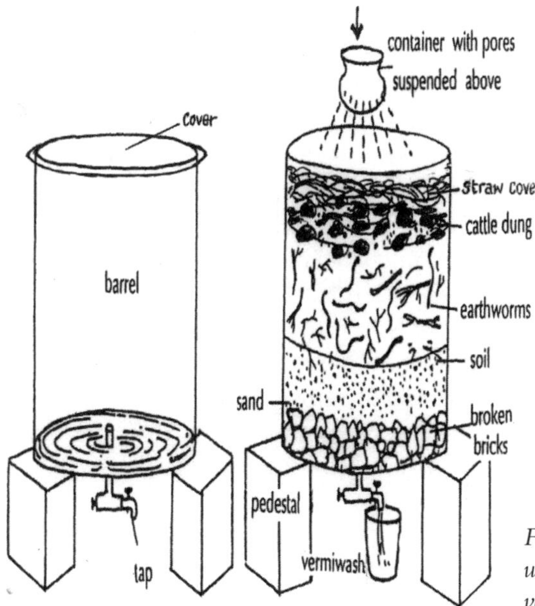

Fig. 3.4 The production of worm compost extract or vermiwash

extract from vermiculture (vermiwash) are highly effective in this respect (Fig 3.4). However, care must be taken to experiment with these materials and avoid applying sprays that have not been sufficiently diluted.

Mention should also be made of activated compost teas applied to the land. These are distinct from the above and are produced by subjecting different kinds of composted material to water extraction and aeration. In addition to nutrients, the teas consist of bacteria, fungi and many other types of organism and therefore potentially help rebalance the organisms of the soil to achieve better breakdown, nutrient retention and control of pathogenic organisms.[31] The technique has had success with improving the soil, especially following years of chemical usage, but one should remember that organisms are more a consequence of good management than the basis of it.

Applying human wastes to the land

When animals digest their food, a certain essence is left over beyond the straightforward life forces it contained. Steiner called this 'ego potentiality'—the result being that animal manure contains this force. In humans, digestion releases these forces to enable us to develop our independent or individual lives. Because *all forces* are withdrawn from what we consume, human waste solids can be regarded as 'dead' matter. Steiner suggested that it might be applied if used in proportion to the number of people living on or working the land. In this case the quantity would not be excessive and some compensation could arise through people's own (astral) influences. Otherwise he counselled that human excreta ought not to be used for human food production. If it is to be applied to the land, then over a period of time 'the proper cycle is from the human being to the plant, from the plant to the animal, from the animal again to the plant, and only then from the plant back to the human being'.[32] The presumption is that humans could become dull and inactive mentally if their food was always grown on soils fertilized with what was once euphemistically called 'night soil'. It is of interest to note that in China's earlier history, while crop production for general consumption was extensively based on use of human excreta, that of the emperor was not; and, of course, even today much use is still made of it. In many other countries its direct use is deemed culturally unacceptable, though this undoubtedly masks the original reason for its rejection.

In Britain and elsewhere, the contemporary need for disposal of sewage sludge to land areas connects with this problem. Although physical nutrients are being returned to the soil together with organic substance which will have certain benefits, crops grown on such soils will have less capacity to support *all* human needs than would be the case with composted farmyard manure or slurry. From the point of view of vital forces, using sewage sludge on land for human food crops is little different from using synthetic chemicals.

The question of human nutrition is tackled in Chapter 10 while the (outer planetary) origin of the forces left over by animals is addressed in Chapter 4.

Wildlife

Wildlife diversity is often taken as a measure of the health of the landscape. One need not be particularly sensitive in order to experience the difference between a farm with natural areas and one without. There is thus a bleakness and monotony without trees, hedgerows and wetland areas. Conversely there can be a feeling of inner nourishment if corridors exist through which wildlife can move and share the landscape with our agriculture.

Most of us are inclined to think of wildlife and its habitats as a separate issue from a working farm landscape. We recognize the role of insects, especially the bee, in pollination, but farmers all too often have cause for a confrontational relationship with nature's wild creatures—from crows to pigeons, and from badgers to foxes. But whether we consider weeds, insects, bird or mammalian pests or plant pathogens, to some degree these questions reflect a failure of agricultural management to achieve a satisfactory ecological balance. Mindful of this, we shall continue the theme of the farm organism by discussing ideas that greatly widen our view of ecology.[33]

Trees

Trees have an impact on landscape in purely aesthetic terms and as regulators of the flow of water and earth energies. Walking through woodlands we become aware of their effects on microclimate while on a larger scale they affect rainfall through their influence on heat and moisture transfer with the atmosphere. The earth's rhythmic exchange of energy is strongly mediated by trees. Schauberger observed energy currents associated with trees, outflowing in the morning and inflowing in the evening, independently corroborating Steiner.[34]

The layer immediately inside the tree's bark is known as the cambium, often equated with 'sap' and regarded by Steiner as a kind of living soil drawn upwards. This consists of upward (xylem) and downward (phloem) movements incorporated into complex neighbouring bundles of cells. These respectively transport water and mineral nutrients in ionic form from the roots to plant extremities, while products of photosynthesis, such as sugars, are taken from leaves to the roots. This has been thought of as a weakly dielectric two-way traffic, positive upwards and negative downwards. The connection between diurnal currents and these plant cellular movements would seem clear. An ancient oak, hollow in the centre but supporting a canopy through survival of its outer layers of bark, illustrates the importance of this living cambial activity (Fig. 3.5).

Fig. 3.5 The living outer bark of an ancient oak tree

The scale of growth of trees makes demands on the life forces of the earth, which become depleted under trees compared with open areas (Fig. 3.6). On the other hand, Steiner indicated that trees have a beneficial effect on a wide area around them, implying

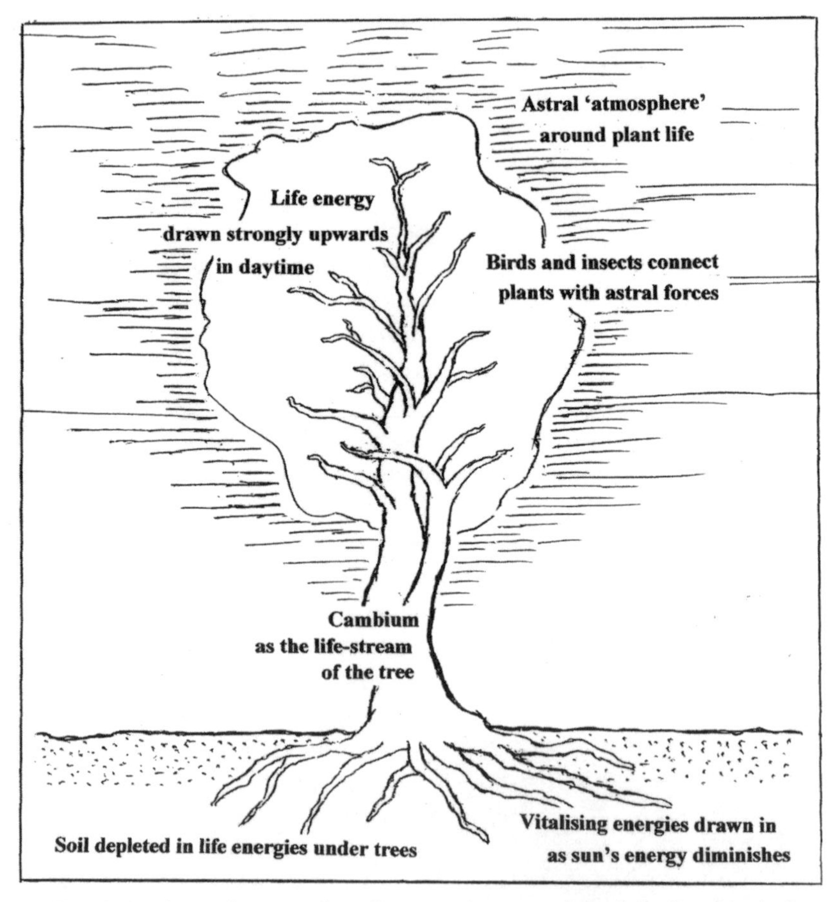

Fig. 3.6 A visualization of astrality around a tree and the daily 'breathing' of life energies

that they act as conduits for vitalizing astral energies. To visualize how these energies may benefit the surroundings it is important not to separate the tree organism and its roots from the soil fabric, from soil organisms and the roots of all other plants in the vicinity. Plants and soil form a unified living environment or continuum, illustrated by the rhizosphere and mycorrhizal network which links plant roots with soil minerals. It has been commented that the largest plant formations on earth are the underground fungal networks of forest ecosystems. The particular energies drawn down by trees thus spread out into all this wider soil life and beyond the limits of the wooded area.

We must visualize the earth's spiritual forces being inspired and vitalized when they expand into the outer atmospheric regions under the sun's influence. We must then visualize the inflow of these forces in the evening and night when they are increasingly ruled by the moon and its special connection with the elements of water and earth. This is the time when alchemical materialization processes used to be most readily accomplished (Chapters 4 and 8). We need to understand that the earth's breathing process is enhanced where there are trees. For all the above reasons, 'alarming' would be an understatement to describe the loss of trees on the planetary scale which has taken place over recent decades. This tragedy is compounded as first one, then another tree species falls prey to pathogens—years ago elms; now horse chestnuts, oaks and others. Fungus disease is identified as a primary cause, but we should begin to understand that if life forces are weakened pathogenic agents are more likely to strike. As such, I believe this to be evidence of a planetary decline in tree vitality.

Before moving away from the subject of trees, it seems that Druids must have recognized the participation of trees in a daily flux of energy when they met in their tree-surrounded sanctuaries, and it is not surprising to learn that the earliest alphabet of the indigenous cultures of western Britain was probably based on tree names.[35] And in ancient times people were aware, as are dowsers, of networks of geodetic energy now known as ley lines. The original meaning of ley was that of a forest glade or clearing, and the first straight prehistoric trackways may well have come about in this way. Such linear or horizontal geomagnetic currents represent an interaction between solar and terrestrial influences, for the energies detected by dowsers vary with the rising or setting sun. While sacred sites were commonly located at their intersections, ultrasonic methods reveal that megalithic sites and standing stones emit high frequency signals at dawn, emissions which are particularly powerful on the mornings of spring and autumn equinox.[36] Some readers may wish to equate such solar-terrestrial energies with the Chinese yang and yin.

Birds and insects

Away from the earth's surface conditions of life change and there is an intensified surrounding envelope of what can be referred to as 'astrality'.[37] This cosmic atmosphere bathes the whole living earth but finds individual expression in the different plant forms. It connects principally with the nitrogen of the atmosphere (Chapters 2 and 4). Those with sensitive vision are able to see such an envelope—as with an aura around human beings. Those of us who cannot still experience its effects, for example flies which hover some distance from the foliage of trees, or even midges which appear to keep up with us on a walk!

The astrality around plants connects them with their cosmic archetypes—their ultimate individuality, source of inspiration or formative influence. The atmosphere's nitrogen provides the physical medium for this to occur. As plants have no consciousness, it is necessary for birds and insects to bring this astral essence into physical contact with them. Flying creatures therefore bestow vitality which otherwise might not occur.[38] So the fact that plants provide major habitat and nourishment for insects and birds actually masks a certain dependence upon these creatures. Probably no one better understood this relationship than the composer Olivier Messiaen, whose work was substantially dedicated to the heavenly qualities of birdsong. Symbiotic relationships occurring within the soil certainly offer a parallel. It is all the more disturbing then, to think how we have reduced bird populations, and how we have treated insects which are attracted to leaf surfaces, buds and flowers. How can nature—or indeed the farm organism—be complete if such intercourse is prevented?

This gesture towards the plant world is wonderfully illustrated by the bee, for aside from pollination bees contribute to greater productivity in areas where they are 'frequent flyers'.[39] Some years ago when visiting villages in Ghana, the writer had cause to compliment several women on their well-kept and productive home gardens. He was surprised when they said they kept bees and as a result usually had better yields than those who did not! As the main subsistence crops were manioc (the yam, cassava) and maize, neither of which require insect pollination, the significance of the above statement was evident. On the other hand, Steiner said, 'Remove these winged creatures [referring to both insects and birds] and the astrality would fail to deliver its true service; and you would soon detect this in a kind of *stunting of the vegetation*.'[40] This underlines the priority we should attach to maintaining bee colonies, for they are so obviously an integral part of the agricultural organism.

Tragically, in many parts of the world, first across Europe and more recently across North America, bees are disappearing. A number of explanations have been offered which quite likely reinforce each other. The depredations of the *Varroa* mite are well known and may reflect a loss of vitality of bees in the face of multiple pesticides. Global warming will be altering the balance of the bees' natural enemies, the incidence of predatory wasps being an example of this. Additionally, we have to account for Colony Collapse Disorder where bees do not return to their hives. Various stress or disorientating factors have been proposed: toxins from GM pollen, the effects of mobile phone networks, WiFi systems and the practice of long-distance transportation of colonies. The latter certainly deserves a fair hearing, for bee colonies have traditionally had a close bond with the bee keeper and are adapted to their particular local area.[41] But after environmental factors have been considered, it seems there are undoubtedly also moral issues to be addressed.[42]

Steiner drew interesting parallels between insect and plant life. The egg and larva, as with the seed and its germination, are much connected to the earth, and if not always *on* the earth certainly are defined by the elements *of* earth and water. Larvae tend to inhabit the soil or bottom of ponds and the watery caterpillar has a parallel with leaf formation. By contrast, the mature insect lives in the air and light, the butterfly classically equating with the flower.[43] So when we next see greenfly on that fresh rose bud we can ponder the image of the insect—a bringer of astrality—attracted to the flower, the part of the plant most strongly connected with the astral realm. For this reason, the arrival of the ant from below or the blue tit from above, creates a picture of an enchanted dinner party! This example, among many others, serves to show the wisdom of biodiversity and the need to preserve it on our farms.

Destructive processes

Healthy plants are less likely to become overwhelmed by insects (except in the case of locust swarms!) or pathogens, whereas the sickly plant will be recognized by its weakened life force and therefore be ripe for destructive forces to take over. In this respect, excess or deficiency of countless growth requirements can render plants vulnerable to disease or predation. Let us consider a particular case discussed by Steiner in the *Agriculture Course.*

In this case, that of fungal disease, the imbalance is between the tendency of water to continually propagate life forces in the soil—connected with lunar influence—and the plant's capacity to progress to maturity through its vegetative stage, to fruit and seed formation. In annuals, a dying process commences around seed formation through *diminished* vitality of the vegetative organs. Paradoxical as it will appear, this process can start to take place at an earlier stage due to *excessive* vitality (lunar influence in the soil due to excess moisture), with the result that plant pathogens are attracted.[44] Counteracting this tendency is an adequacy of light and warmth which forms the basis for counter-measures discussed in Chapter 5.

If we consider the proportions of micro-organisms in the soils of different ecosystems, in immature, skeletal soils we tend to find only bacteria, while healthy arable soils have a fungal biomass about equal to that of bacteria. The proportion of fungal biomass increases in grassland, especially old meadows, where it may be 2–3 times that of bacteria. We should bear this in mind when considering the contribution of leys to farm rotations. Here, *Streptomyces* (a filamentous bacterium, an actinomycete) releases metabolites called geosmins that are said to give soil its 'earthy' odour. Finally, under deciduous and coniferous woodland the biomass of fungal organisms is 50 and 200 times greater than that of bacteria, the primary decomposers of woody material being insects (mainly ants) and fungi.

Steiner urged us to be aware that soil fungi were of great importance in bringing about balance on the farm. What he wanted us to consider was that for all the biological produce we desire to achieve, the landscape must somehow harmonize this by engaging its destructive forces. If this is not possible—if we so intensify and maximize the growing area so that there is little scope for fungal activity, then the stage is set for pathogenic organisms to attack our growing crops.[45] The advance of pest attack following agricultural intensification (and a parallel situation with animals) connects with this idea and underlines the fact that we can rightly talk of a farm organism becoming sick.

So whether one considers trees and other wild flora or birds, invertebrates, insects and soil organisms, one cannot escape the conclusion that the fruitfulness of the earth depends on its wildlife—and on us sharing the earth with other creatures. This might well be expressed in the axiom *allow nature her place and our best interests will be served.*

The dissipation of life

We must finally try to understand that at the end of life there is a subtle interchange between the earth, on the one hand, and the surrounding universe which has brought about that life in the first place. Here, we are not concerned with the force that has held life together but of something more fundamental. What has been built up in a living *form* on the earth must be able to dissipate. The *form* is what archetypal forces (or spiritual blueprints) have built into living carbon structures. In such a structure, the spiritual[46] has become physical. At the end of life, this spiritual principle must be led back to the cosmos. This takes place through the medium of the least physical or most spiritual of earthly substances—hydrogen.[47] We can imagine that the initial stages of this process are accomplished through decomposition, while hydrogen's eventual release may well relate to movements of water in the atmosphere and the dissociation of hydrogen and oxygen in the upper atmosphere where hydrogen is constantly escaping. We can try to visualize—as with water and homoeopathy—that hydrogen carries the memory of the forms created.

Steiner referred also to a different aspect of this dying process. He said that birds, when they die, have the role of carrying away a spiritual essence of earthly substance. They have 'the task in cosmic existence of spiritualizing earthly matter', and of 'transferring this spiritualized earthly matter to the universe'.[48] Note that here we are talking of *earthly matter* rather than *archetypal forces and the forms they have produced*. He goes on to say that the butterfly achieves this process to a greater degree than the bird, for 'even in life it is continually restoring spiritualized matter to the cosmic environment of the earth'.

In both these cases the raying back of forces from the living world provides for the spiritual world a vision of what has been created on the earth. Elsewhere Steiner indicates that such a process will happen in the far future of Earth evolution, for the earth as we know it has only come about through the progressive 'condensation' of spiritual essence, and will eventually be returned to that state.

Concluding remarks

Having considered different aspects of the farm organism, we should recall that at the outset, our ecologically based model was essentially two-dimensional. Together with the picture of cosmic influences from Chapter 2, the further elements presented above give more than enough justification for embracing a third dimension. Self-evidently, the farm organism includes not only the soil and earth below but the wide universe above. The various preparations to be discussed in Chapter 5 are tools for strengthening this organism.

We will be familiar with the hydrological cycle in which water is evaporated, rises in the atmosphere, condenses and falls as rain. So too, the farm must be considered as involved in a cycle of life. It receives influences from the cosmos, participates in creative processes and finally, through various agencies, experiences its dissolution. There is in this, a real feeling of interplay between cosmos and earth—that they really do form a unity.

Nevertheless, what does or does not constitute a farm organism in practice is a matter for each individual to decide or, indeed, to experience. This helps us appreciate that what makes a farm is not simply the interplay of physical and biological resources, or even their efficient use, but the direction given to the whole process by the farmer or group of people leading its activities. In this respect each farm has what we may call an *individuality*, something greater than the sum of its visible parts. Earlier, we sketched the relationship of humans to the entire earth. Now we can say that the farmer has the potential to draw together all the elements both cultivated and wild into an integrated whole (Fig. 3.7). The attention given to the various aspects discussed in this chapter create a basis for distinctive local produce of recognized quality in the marketplace.

We commented on the fact that many farms fall short of established ideals. In this connection it is important to realize that the human element is of prime importance when working with nature, as there are beings—nature spirits and elemental beings—closely associated with every aspect of the organic world, in fact with all states of matter.[49] While the elemental beings connected with the life of plants are referred to in Chapter 4, we must appreciate that a whole hierarchy of beings works with nature's processes on different levels. Lesser beings connected with the life of trees are thus guided in their work by tree spirits or devas. The picture relating to the striving of human beings on farms and in other institutions of life is no less awesome, for beings of the angelic order watch over us and guide people in many invisible ways. Think about this when next you remember something vital 'just in time' or have an important chance meeting with someone which could all too easily be described as 'pure coincidence'!

Fig. 3.7 An image of the human being in relation to a farm (original drawing by F. Schikorr)

We can therefore understand why a positive attitude driven by a sense of purpose is so important, for it enables higher beings to help us. Human beings today sense they are alone in the world. These final remarks are simply to say that in fact we are not alone in what we do.

4. The Working of Cosmic Energies in Plant and Soil

For life on earth we take for granted the importance of the sun—to a lesser extent the moon—but the great majority of people now feel little connection with what lies out in the cosmos (aside from surreptitious glances at the horoscope!). But since the most ancient times people have been aware of cosmic influences on life—and not just as primitive superstition. In this respect the present world view, reflected in our whole culture and educational system, illustrates a consciousness that has fundamentally changed in the past few hundred years. Current scientific thought accepts the cosmos (= 'beautiful ordering' in the original Greek) as a source of energy, but what it cannot comprehend is that such energies carry more complex and qualitative essences which underpin life on earth.

In this chapter, using the guidance of Rudolf Steiner and the experiences of others, I want to show how cosmic forces work in the realm of plant and soil. This is not just about how light directly affects the green plant but how in a broader sense it is absorbed by the earth, and how this absorbed energy contributes to the growth of plants and to nutrition more widely.

The green plant and the nature of light

In addition to physical properties, the rays of the sun also possess etheric qualities.[1] Visible light comprises only a limited part of the electromagnetic spectrum (Fig. 2.4). While the warmth of the atmosphere depends on solar short-wave radiation being re-radiated from surfaces as long-wave radiation, the vital process of photosynthesis is triggered by certain bands of direct solar radiation. The most intense part of the solar beam falls in the green part of the spectrum. Paradoxically, the chlorophyll pigment ensures that much sunlight is *reflected*, which we see as green. In fact chlorophyll absorbs mainly violet, blue, orange and red—very little green. Other pigments control the absorption of further wavebands.[2] In photosynthesis, light energy is fixed as energy contained in matter—as sunlight and carbon dioxide yield the sugar glucose. This in turn, becomes the basic structural unit for the carbohydrates, starch and cellulose—an amazingly complex series of steps if one examines the process biochemically. Furthermore, variations in *light intensity* and *photoperiod* underlie important differences in photosynthetic pathways (C3 and C4 plants) and also trigger flower formation.

But the process of synthesis in the leaf cannot take place without the participation of an upward flow of water and nutrients from plant roots. Substances are required for osmotic balance, as enzymes or precursors. Potassium identifies itself with this upward stream and with the passage of salts across cell membranes while iron, as part of the haem structure, is required as a precursor to the formation of chlorophyll molecules.[3] Of paramount importance, also, is phosphorus. Through such complex molecules as NADPH and ATP, phosphorus causes excitation of chlorophyll subtypes to produce glucose. It is uniquely endowed with this capability because of its intimate connection to the light and warmth ethers (Chapter 5). Thus solar energy is transferred into the energy carried within physical substance.

Yet if the sun is as important a force as we know it to be, it seems strange that so significant a part of its spectral energy is reflected. Here, we should recall the distinct roles of the inner and outer planets as outlined in Chapter 2. Chemical and biochemical processes are clearly mediated by the *inner* planets. Certain bands of short-wave radiation are absorbed because they are the ones required to activate and accomplish the process of photosynthesis.

In addition, the sun conducts forces to us that are mediated by the *outer* planets. Because these are etheric in nature and create the outer forms of the plant, we can assume they will not be subject to laws that govern the passage of electromagnetic radiations. Such a 'forming or shaping principle' must also operate simultaneously with photosynthesis. It would therefore appear that the inner essence of sunlight is 'absorbed', but is it possible to find evidence to support this idea?

Discovering a signature for the light ether

Whenever we discuss light in a prosaic way, we need to pause and reflect that, after all, it constitutes cosmic energy and originates from the sacrifice of spiritual beings. Furthermore, in Chapter 2, we related crystalline structures to archetypal cosmic forces. Strongly connected with the element of light is silica (quartz, silicon dioxide). Its crystals display equiangular six-sidedness, arguably closest of any straight-sided shape to a circular 'cosmic' form (Figs 2.2 and 4.1). In fact, using traditionally adopted symbols, this form unites warmth and light ethers, for within the circle (warmth) can be fitted two interpenetrating triangles (light)—a hexagon being the result (Fig. 4.2). Such forms are found in the organic world. In the beehive we see hexagonal forms in the honeycomb (Fig. 4.3). The bee

Fig. 4.1 Prismatic crystals of quartz

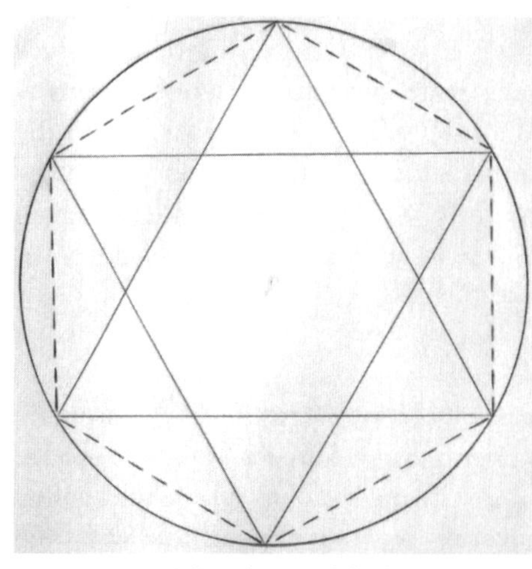

Fig. 4.2 The relation of the hexagon to cosmic ethers

is said to have a strong relationship to the sun and is 'formed by the same forces that in the earth form quartz'.[4] We should also note that the compound eye of the bee and other insects is composed of hundreds of tiny hexagonal lenses. In this way we begin to see in six-sidedness a common formative principle or signature of the light ether—a creative force contained within physical sunlight.

If we now examine the chlorophyll structure (Fig. 4.4)—a porphyrin, similar to haemoglobin—we see four six-sided ring structures held dynamically by a central magnesium ion. Magnesium, like potassium, is a highly reactive, lime-related substance which has risen up the plant. Photosynthesis can therefore be pictured as a marriage between cosmic light and earthly substance. So here, displayed in just one of an infinite variety of organic molecules, we see the six-sided structural form. It is, of course, the particular chemistry of carbon which permits this, yet carbon manifests in other structures besides. Forces within the light lie behind chemical structures, while simultaneously these structures become the means by which a plant grows. In other words, light forces hold archetypal forms in place while the world of water and chemistry works to fill the forms with substance (see the role of elemental beings below).

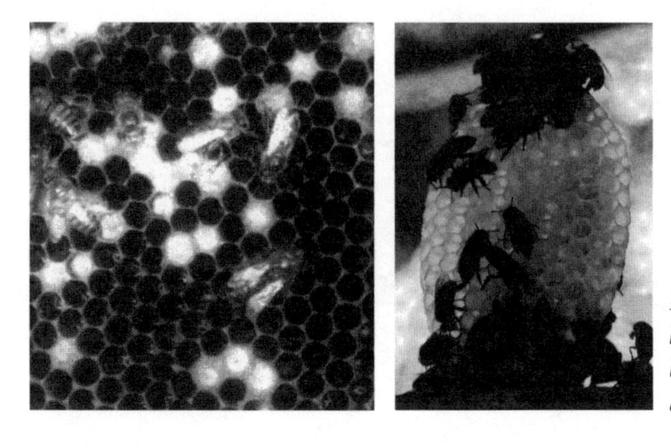

Fig. 4.3 The hexagonal cells of beehives with bees active

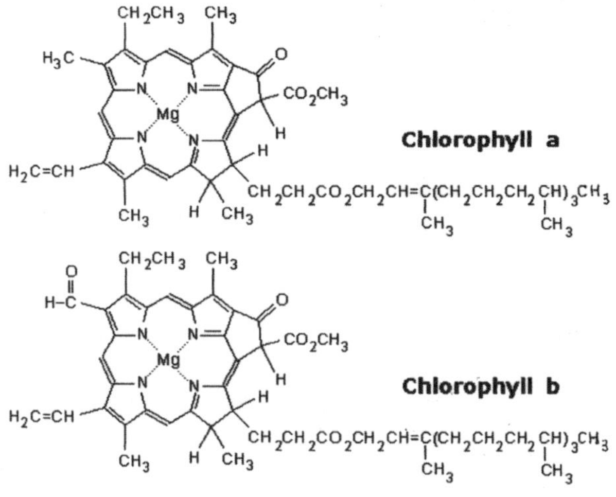

Chlorophyll a

CH_2CH_3 CH_3 H_3C O CO_2CH_3 N Mg H $H_2C{=}CH$ CH_3 H CH_3 $CH_2CH_2CO_2CH_2CH{=}C(CH_2CH_2CH_2CH)_3CH_3$ CH_3 CH_3

Chlorophyll b

CH_2CH_3 CH_3 O H–C O CO_2CH_3 N Mg H $H_2C{=}CH$ CH_3 H CH_3 $CH_2CH_2CO_2CH_2CH{=}C(CH_2CH_2CH_2CH)_3CH_3$ CH_3 CH_3

Fig. 4.4 Structure of the two most common chlorophylls

In the chlorophyll structure each of the heterocyclic rings (ones made up of different atoms) is three-dimensional, like doughnuts, with electronic clouds bounding *inner spaces*. The double lines (double bonds) denote that individual 'cells' of the molecule are deficient in electrons. As a result, electrons have to switch their positions at immense speed. This is a common occurrence in the ring structures of organic chemistry and is practically like setting up a vortex within individual cells of the structure. Just as living organisms separate themselves off from an exterior environment, we should take particular note of these interior spaces. They are centres of receptive chaos in which creative or transformative processes can take place.[5] Likewise, one might think of inner development characterizing the cells of a beehive, of the root cells modified by vesicular-arbuscular mycorrhizae, and certainly of the creative process within the mammalian womb.

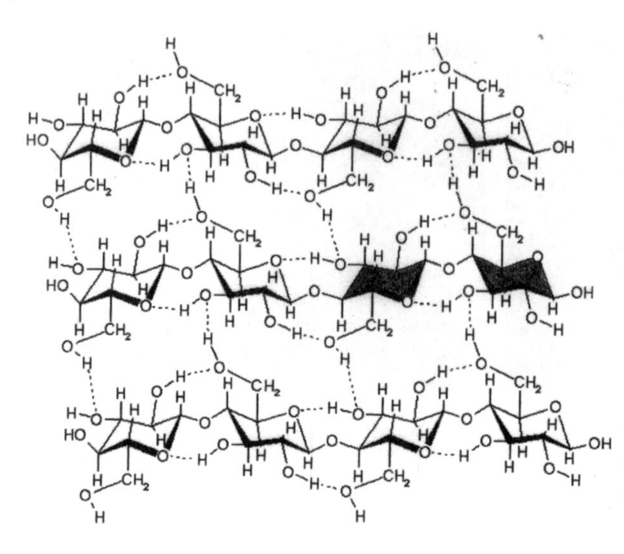

Fig. 4.5 The structure of cellulose. Two of the six-membered rings have been shaded to help recognition. Carbon atoms are unmarked and lie at the junction of bonds, indicated by lines

But considering the plant as a whole, at least half of its total mass is composed of the carbohydrate we know as cellulose (Fig. 4.5). Fundamentally a polymerized sugar, or polysaccharide, we should note that this too is based on six-membered rings. The arrangement of these 'rings' is described as puckered owing to the geometry of the bonds between adjacent atoms. Note how crucial oxygen is in the coherence of the structure, for it is this which carries the etheric currents. Chemists actually refer to oxygens which bridge carbon structures as ether linkages! Adjacent units are offset as indicated by the alternate positions of these linking oxygen atoms.[6] This is an ideal arrangement for achieving rigid fibrils for cell walls. These linear molecules lie close together and because of abundant hydroxyl (—OH) groups, the individual chains can be linked by hydrogen bonds.

It is now possible to sense a consistency of molecular patterning or formative influence so that chlorophyll and chloroplasts can be viewed as *gateways for the light into the being of a plant* while carbohydrates can be visualized as *crystallized sunlight.* In fact, we may begin to feel happier with the idea that physically manifest light is merely the outer garment for the light and warmth ethers, a multifaceted spiritual stream incorporating different types of creative impulse.

Light as formative and inspirational force

It will be evident that nature's created forms are amazingly diverse above ground—in the light—yet are relatively simple below ground and in the ocean depths where light is lacking. Consider the formation of eyes.

Without light, other sensory mechanisms develop, and without light, the eye degenerates, as with pit ponies of earlier coal mining days. Eyes were formed by ether forces within the light so that it was possible to experience the illuminated world provided by that light.

The leaves of different plants display an enormous variety of forms which relate to the relative influence of a 'lime or silica principle'—again related to the influence of inner or outer planets. Leaf forms may thus relate more to light and warmth (serrated and spikier forms, silica) or to the water and earth elements (rounder forms, lime). One can observe these principles when comparing the leaves of dandelions growing in a shady place (rounder) with those growing in the open (spikier). Many herbs display leaf metamorphosis, being characterized by rounder leaf shapes nearer to the ground (as with cotyledon leaves), these becoming more pointed as one approaches the flower (Fig. 4.6).[7] It is in the flowers that what was referred to as 'astrality' (Chapters 2 and 3) works itself most strongly into the plant.

The soil environment is *without* light and, while not without diversity of forms, growth or movement is constrained by the size of soil pores. There is also no evidence of metamorphosis. With roots, elongation and ramification form a largely repetitive, dendritic or fractal-type pattern (Fig. 1.11). Many soil organisms vary in size rather than design and, aside from surface-dwellers, the bulk of the population is built on a single cell, linear (tubular) or branching principle.

How does light affect the human being? The first thing to realize is that it penetrates all our senses, not just vision.[8] As we shall see in Chapter 10, part of our nutrition is 'cosmic', the most obvious example being the effect of sunlight on production of vitamin D. Phosphorus, as we have

Fig. 4.6 Leaf metamorphosis. Example of Geum urbanum *(herb bennet or wood avens)*

noted, has a particular relationship to the warmth and light ethers, and can be considered the physical bearer of these. It is to be found in every cell of our body and is second only to calcium as a constituent of our bones and teeth. Light in a wider sense touches us in the realm of thought. Lehrs, developing an insight of Goethe, stated that it is really the light ether which enables us to form objective conceptions about the natural world.[9] We speak of 'seeing the light' in the sense of understanding the truth about things, and in John's Gospel Christ says, 'I am the *light* of the world.' This is territory beyond our present scope but it illustrates the fact that 'light' is more than simply an aid to physical vision.

Nature spirits and elemental beings

In Chapter 2 we indicated that groups of spiritual beings are involved with creative processes. In relation to the foregoing, it is interesting to note that overseeing the life of plants are beings at the level of archangels, *beings of light.*[10] But how are divine spiritual beings actually able to function in the physical world, consisting of its solid, liquid and airy elements? Rudolf Steiner's response was: 'They send down elemental beings; they imprison them in air, water and earth. These are the elemental messengers of the spiritual, creative, formative beings.'[11] Elemental beings are therefore enchanted in everything surrounding us! 'Beings of light' do not, it seems, expect their influence to be conducted in an inanimate fashion, as might have been implied in the above discussion of photosynthesis!

We should recall from Chapter 3 that all the earth's 'living processes', atmospheric and oceanic, are similarly mediated. These nature beings connect ultimately with the earth's living or etheric organism, and are therefore influenced by its diurnal and seasonal rhythms.[12] In view of the difficulty of thinking of the earth as a living entity, we should realize that while 21 per cent of the earth's atmosphere is oxygen its crustal layers comprise 46 per cent by weight of this element and an astonishing 92 per cent by volume. Oxygen, as will be commented on later, is the physical bearer of etheric or life forces.

Four types of elemental being are connected with the primal or Aristotelian elements and work in relation to different parts of the plant. These are: the gnomes in the root realm; the undines (nymphs) within the water movements of the plant, in stem and leaves; the sylphs in the light element of leaf and flowering processes; and finally, fire spirits in the warmth element, connecting with fruit and seed formation. So within the 'astral' envelope around the plant, elementals form leaves and flowers according to plans determined by archetypal cosmic forces. In the words of Steiner 'with the help of up-streaming substances worked on by undines, sylphs weave an ideal plant form out of the light'.[13] This reinforces the idea of the plant representing a 'marriage' between earthly and cosmic forces.

All these beings have at some time been seen by individuals, some with special 'gifts', some who have trained themselves, and by no means infrequently by children. These and other nature spirits are documented in the folklore of different cultures, in the writings of Paracelsus, and appear in Shakespeare's *The Tempest.*[14] We shall return later to consider the particular contributions of these beings.

The passage of 'light' into the earth

Energized and supported by the sun though plants are, they are also dependant on *forces working from below upwards*. Steiner informed us that 'what comes from the sun in conjunction with the *outer planets* is absorbed by the ground and then radiated upwards'.[15]

So how are cosmic forces absorbed by the earth in the first place? To begin with, let us dispel certain possibilities! Visible light hardly penetrates more than a few millimetres of soil while very short wavelength radiation, like gamma and cosmic rays, can penetrate the ground but, thanks to the earth's magnetosphere, is of low incidence. On the other hand, we know from seismology that long waves of differing waveform from earthquakes move unimpeded through the earth's crust. Since the planets are low-temperature bodies, we might imagine that radiation from them would tend to be of longer wavelength—like sound. Might we therefore be dealing with information carried by cosmically generated long waves? Such lines of thought merely result in endless speculation and it seems better to accept, as we did in Chapter 2, that while various forms of energy may *carry* information they do not constitute the message itself. There are, however, clues as to how we should be thinking. The absorption of these solar and outer planetary forces is first and foremost promoted by the *silica content* of the earth—silica as sandstones and silicate rock minerals. In Chapter 2 we referred to silica as having the character of a universal 'cosmic sense', so it is possible to see this role being fulfilled.

In the *Agriculture Course* Steiner refers to water and solids being 'alive' above ground and 'dead' below—whereas for light, air and warmth, the reverse is the case.[16] What did he mean by this? My assumption is that it signals a polarity in the working of life forces in conjunction with the inner and outer planets. For this reason, nitrogen in the air is referred to as 'dead' while it becomes 'alive' within the earth. Such animation of substance takes place where appropriate *etheric or life forces* prevail. Nitrogen and oxygen thus become 'alive' inside the human being. In this respect, in addition to actual living organisms and plant roots, soil possesses a dispersed or uncombined life force from decomposed matter. Thus, when cosmic forces engage with the earth they do so not with the purely physical substances of atmosphere and earth but with the earth's etheric organism, a notable contributor to which is oxygen. Therefore, when Steiner spoke of the important role of *silica* in drawing in these influences it is likely he saw silicon's connection with oxygen as a crucial part of the picture.

The presumption is that within an etheric environment cosmic forces release their form and life-creating potential, so to think of them as having to penetrate a barrier of physical substance will not be helpful to our understanding. Whether one considers the soil or the plants growing in it, we are thus involved with the realm of life and cannot be limited by laws which affect non-living physical substance. Indeed, conditions within the soil change *etherically* with the passage of the seasons, and life forces are intensified over the winter months. There is a time between mid-January and mid-February (northern hemisphere) when 'forces of crystallization' are at their maximum.[17] Our thoughts about the working of these solar and outer planetary forces must therefore take the earth's seasonal (and diurnal) 'breathing processes' into account, so let us take a moment to ponder Rudolf Steiner's meditation:

The earth's soul sleeps
in summer's heat;
then the sun's mirror blazes
in the outer world.

The earth's soul wakes
in winter's cold;
then the true sun shines
spiritually within.

In summer's joyful day
earth sleeps deep;
in winter's holy night
earth rouses, wakes.

Earlier we noted that plants are composed of much cellulose which bears within it the signature of a creative light force or light ether. We may therefore visualize a communication within the cell structure of plants which brings the world of light down to the roots. No doubt both the silica and phosphorus content of cells plays a part here. Steiner refers splendidly to such a process when he says that root spirits (or gnomes) receive their ideas from 'what trickles down to them' through plants from above. 'Especially from summer to autumn these spiritual essences are received and distributed within the earth by these gnomes.'[18] This would indicate that 'light' is contained within the etheric organism of the plant and that when vegetative life fades it drains into the soil.

Steiner elsewhere states that the light and warmth from one summer continues within the earth, forming the basis for the following year's growth.[19] At this point we should refer to an experiment by Lili Kolisko.[20] Growing plants at various depths *beneath* the ground surface, she not only detected that the moon's phases were registered by variations in growth rate but that the course of the growing season was progressively retarded the greater the depth. While this shows that cosmic influences do indeed penetrate the earth, it also seems that the weaker the penetration of light forces the more retarded the upward forces of growth the following year. Prehistoric 'chambered tombs' also provide insight regarding the penetration of 'light' into the earth. Originally to facilitate initiation into the mysteries, those involved were cut off from the outer world, only experiencing *the inner or spiritual light* of the sun.[21]

It would therefore appear that light forces penetrate soil and plants via etheric pathways. As these forces connect strongly with daytime and summer, they are clearly subject to the earth's breathing rhythms. Furthermore, whether they penetrate by way of soil fabric or plant life, the combination of silicon and oxygen plays a key role in their reception. We need now to explore how such forces are held in the soil, and therefore whether they are always held there to the same extent.

Retention of cosmic forces in the soil

Solar and outer planetary influences—the light forces already mentioned—are required to infuse earthly matter. The reason for this is that physical life is underpinned by its own ether force. Matter (i.e. chemical substances) must therefore bear this force—the life ether—in order to fully support the healthy growth of plants, and subsequently animals and human beings. These 'light' forces must therefore be brought into such a relationship with the soil that a *qualitative transformation of substance* can occur. Plant nutrients will then be able to carry this force into the life processes of the plant.

In contrasting the roles of silica and lime, Steiner indicated that clay mediates between them and is connected to what he called the 'cosmic upward stream'. He also stated that '*substances like clay* support the upward flow of the effects of the cosmic entities in the ground'. In this regard, we should point out that both humus and clay are colloids, having neither the properties of true solids nor of solutions. This needs to be remembered when considering Steiner's ubiquitous use of the term 'clay', for the intimate association between these two differently formed substances urges a broader interpretation of Steiner's terminology.[22]

While silica and alumina are the principal components of clay structures, lime and kindred substances are held in the soil by a combination of clay and humus. So in terms of physical matter, the 'upward stream' involves calcium and other nutrient substances dissolved in water. But referring to it as 'cosmic' implies that in addition to substance *force* is involved—the life ether. Our next task is to understand why clay, or rather colloidal materials, should be key to the retention of this cosmic light essence within the soil.

The nature of clays

The mineral part of soil consists of sand, silt and clay fractions, clay referring to the finest soil particles. But the clay fraction may comprise different minerals, most important of which are alumino-silicates which come in various generic forms. Aluminium and silicon are amphoteric, having the propensity to form both acids and bases. As a universal component of clay minerals, aluminium, together with phosphorus, exercises a mediating role between formative forces brought in by silica and the substance-building activity of the lime family.[23]

It is characteristic of clays that they have colloidal properties due to their minute particle size—in other words they disperse evenly in water (as with milk), and instead of settling due to gravity, they remain suspended. Individual crystals (or flakes) of alumino-silicate clay vary from $0.01–5 \times 10^{-3}$ mm, much smaller than the finest sand at around 0.8×10^{-1} mm. All clay crystals have a lattice-type structure and, while identifiably crystalline, are not bounded by solid edges. They might be likened to the skeleton of a modern building without its exterior cladding, except that the height between individual 'floors' is only $0.5–1.5 \times 10^{-8}$ mm! The exterior surfaces of these structures are normally slightly electronegative while calcium and its allies, such as magnesium and potassium, are electropositive. The result is that a cloud of these positively charged ions (cations) of diameters ranging from $1.0–2.5 \times 10^{-9}$ mm will tend to surround crystal surfaces like bees trying to enter a hive. On the edges of the crystals the situation is usually the opposite, the mineral presenting a net positive charge which attracts negatively charged substances (anions) like phosphate.

Fig. 4.7 Stacking of a pure expanded clay mineral. Diameter of platelets c. *0.2 μm. (Queensland University of Technology)*

Clay minerals have mostly come about through the alteration of crystalline rocks and in contrast to everyone's favourite quartz crystals, which are hard and remain fairly stable, their structures are subject to constant change according to the exterior weathering environment. Some undergo great expansion, appearing like concertinas or book pages, as we see in Fig. 4.7.

Examining the lattice of the simplest clay mineral, kaolinite, in *sectional view* (Fig. 4.8) we find layers of aluminium and silicon ions with oxygen and hydroxyl ions arranged around them. This gives rise to polyhedral, platelike crystals as seen in Fig. 4.9. If we proceed to observe the kaolinite crystal in *plan view* (Fig. 4.10) we see at the molecular level what comes about through the linking of silica tetrahedra—it is the hexagonal pattern which expresses itself in the whole crystal.

As this is a scale representation, it will be noticed how much internal space there is in clay minerals—space to which we drew attention earlier. Indeed, if we return to Fig. 4.8, the linkages between units of the crystal

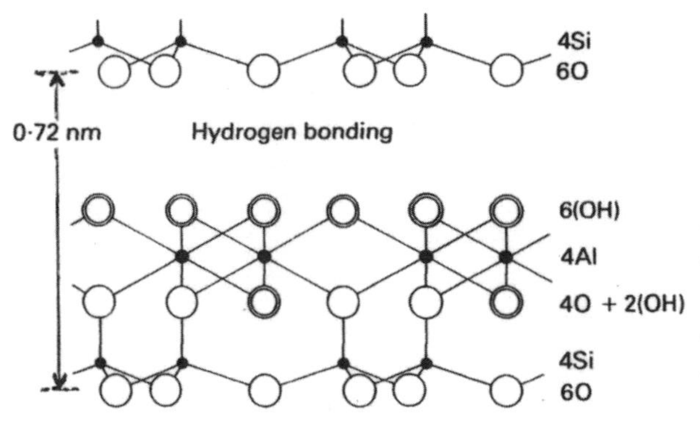

Fig. 4.8 Sectional arrangement of ions in the clay mineral kaolinite (from E.A. Fitzpatrick, Soils, *1980)*

Fig. 4.9 Hexagonal kaolinite crystals. Diameter c. *10–15 μm. (University of New Mexico)*

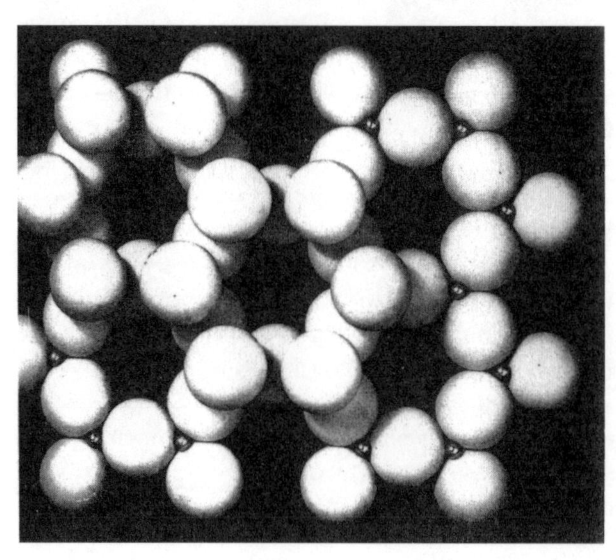

Fig. 4.10 Model of silicon-oxygen tetrahedral sheet with some oxygen ions removed to show the smaller silicons (from Fitzpatrick, 1980)

are more tenuous where hydrogen bonding occurs. Here, there is a great deal of inter-ionic space which in some clays leads to water penetrating the lattice. If we explore the limits of interior space we can say that while an ion may be portrayed as a spherical phenomenon, in reality the electrons defining it occupy virtually no space at all.

The importance of soils with a living quality

To penetrate earthly matter and infuse it with a life ether force, Steiner emphasized that a 'living soil' is required. So what exactly is meant by a *living soil* in this context? Surely we already know the answer to this from

Chapter 1. A living soil is one with living plants and organisms. It predicates organic management with the use of manures and compost.

But associated with this living and decaying soil life there are, as we said in Chapter 3, radiating forces present within the soil. Composted matter thus permeates the mineral fabric of soil with an essence of life—unbound soil life we might call it—and by way of its nitrogen content a sensitivity to cosmic forces. Steiner made reference to this when he said that 'in manuring we must bring to the earth kingdom enough nitrogen *to carry the living property to those structures in the earth-kingdom to which it must be carried*—under the plant', where the soil is.[24] So, once again it is the 'sensitivity' of nitrogen which guides these ether forces into the deeper recesses of the soil fabric. In this respect, a chemically treated soil with a lower level of life activity will be disadvantaged and relatively 'deaf' to cosmic signals.

A further thought could be added. Atmospheric nitrogen was referred to as becoming 'alive' when it was in the soil. It follows that the greater the life quality of the soil the more animated this nitrogen is likely to become. This indicates the importance of good soil aeration for the working of these processes.

Clay and humus complexes

A living soil also implies a continuing supply of organic materials which, through the action of micro-organisms, are gradually transformed into humus (*humus*, in Latin, had the broader meaning of soil, earth or the ground). In most soils humus will form complexes with clay minerals. These come about through attraction of oppositely charged parts of the structures or through divalent cations such as calcium, forming bridges between them. The complex greatly increases the ability of soil to hold water and substances involved in plant mineral nutrition. It also increases the longevity of the humus, for, as growers well know, where clay content is low it is always a struggle to maintain satisfactory organic matter levels. In sandy soils, humus is indispensable as a basis of fertility, while in many tropical soils with a high content of iron and aluminium oxides, to maintain humus content is the key to continued production.

Humus itself is unidentifiable as to its plant or animal origin and represents organic matter that has passed over from the living to the mineral realm. It is the most stable form of organic carbon in the soil, but even so it needs constant replenishment. Its typical molecular structure, that of a polymer (Fig. 4.11), though in no way susceptible to categorization, as with clays, still bears a resemblance to the cellulosic structure discussed earlier. Is it unreasonable to suggest that structures originally

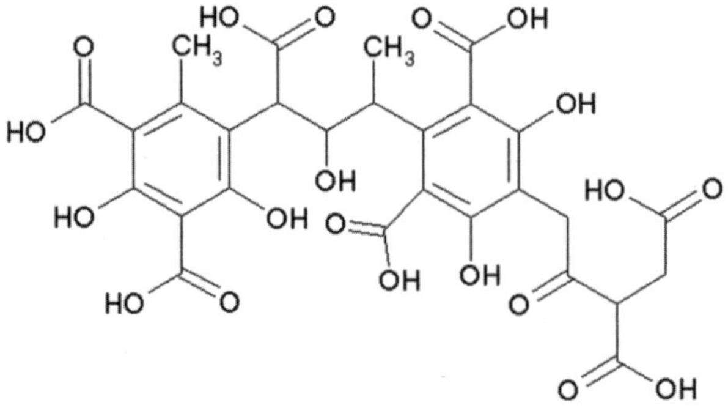

Fig. 4.11 Generalized representation of part of a humus molecule. Three-dimensional character is present but cannot be shown here

formed through the action of light—and now humus—should not also be receptive to formative forces carried by light?[25]

The humus-clay association would appear then to be a vessel for the concentration of cosmic-etheric activity within the soil, which in the wintertime of temperate regions behaves like a battery on charge. After all, oxygen is a key component of clay minerals and we continue to emphasize its importance as a 'carrier of the etheric'. Furthermore, calcium and other cations held around these tiny particles—or bridging between them—should be the prime beneficiaries of etheric forces concentrated within the interior spaces. A recent communication with nature spirits illustrates not only the importance of the interior voids within clay but also mediation of the entire process by elemental beings of the earth. To the question of how cosmic forces are held in the earth, the reply came: '*We* (the gnomes) *preserve them in the structures of the crystals . . . there is something like a matrix in crystals and the cosmic forces seep into this matrix. It looks very beautiful.*'[26] One might visualize minute vortices within the voids of the clay lattice and other suitable interior spaces.

The period in which such processes take place will vary in different parts of the world. The seasonal rhythm of northern lands involves a main inflow of such forces from late summer onwards and a withdrawal through plant growth the following year. In the tropics, the dynamic which maintains the momentum of life is the strong diurnal rhythm. In this way the humus-clay complex gathers energy and has special effects on the chemical substances drawn to its influence—a process facilitated by soils with vitality. Substances held in such soils will carry an added force as they are subsequently drawn into the stream of nutrients for plant growth.

We shall follow the path of these plant nutrients in the final part of this chapter after exploring the nature and relationships of the elements themselves.

The origin and role of chemical elements

The arrangement of elements according to their mass and electronic characteristics is known as the periodic table (Fig 4.12). This highlights resemblances in chemical properties and offers us a tool for predicting each element's properties. One of the first things to notice is that hydrogen, carbon, oxygen and nitrogen, among the lightest elements, have an atmospheric affinity and comprise over 95% of plant dry matter (see below).[27] While carbon, hydrogen and oxygen are the main con-

A	B	C Transition elements										D	E	F	G	H	I
1 H 1·008																	2 He 4·003
3 Li 6·941	4 Be 9·012											5 B 10·81	6 C 12·011	7 N 14·007	8 O 15·999	9 F 18·998	10 Ne 20·179
11 Na 22·990	12 Mg 24·305											13 Al 26·982	14 Si 28·086	15 P 30·974	16 S 32·06	17 Cl 35·453	18 Ar 39·948
19 K 39·098	20 Ca 40·08	21 Sc 44·956	22 Ti 47·90	23 V 50·941	24 Cr 51·996	25 Mn 54·938	26 Fe 55·847	27 Co 58·933	28 Ni 58·70	29 Cu 63·546	30 Zn 65·38	31 Ga 69·72	32 Ge 72·59	33 As 74·922	34 Se 78·96	35 Br 79·904	36 Kr 83·80
37 Rb 85·468	38 Sr 87·62	39 Y 88·906	40 Zr 91·22	41 Nb 92·906	42 Mo 95·94	43 Tc (97)	44 Ru 101·07	45 Rh 102·906	46 Pd 106·4	47 Ag 107·868	48 Cd 112·40	49 In 114·82	50 Sn 118·69	51 Sb 121·75	52 Te 127·60	53 I 126·905	54 Xe 131·30
55 Cs 132·905	56 Ba 137·34	57 La 138·906	72 Hf 178·49	73 Ta 180·948	74 W 183·85	75 Re 186·207	76 Os 192·22	77 Ir 192·44	78 Pt 195·09	79 Au 196·966	80 Hg 200·59	81 Tl 204·37	82 Pb 207·2	83 Bi 208·980	84 Po (209)	85 At (210)	86 Rn (222)

Fig. 4.12 The periodic table of elements. A. Group 1 (Hydrogen and the alkali metals). B. Group 2 (Alkaline earths). C. Metallic substances classed in various groups. D. Group 3. E. Group 4. F. Group 5. G. Group 6. H. Group 7 (Halogens). I. Group 8 (Inert gases). Only elements to the end of period 6 (atomic number 86) are shown. The lower number in each cell is the average atomic weight of each element compared to $\frac{1}{12}$ of the weight of the isotope ^{12}C

Typical % macro and micronutrient content of plant dry matter

Carbon	C	44.5	Sulphur	S	0.1
Oxygen	O	44.5	Iron	Fe	0.01
Hydrogen	H	6.0	Manganese	Mn	0.008
Nitrogen	N	2.0	Zinc	Zn	0.004
Potassium	K	1.5	Boron	B	0.003
Calcium	Ca	0.8	Copper	Cu	0.0005
Phosphorus	P	0.2	Molybdenum	Mo	0.0001
Magnesium	Mg	0.2	Cobalt	Co	0.00005

stituents of cellulose, proteins and oils depend on nitrogen and sulphur in addition. Why are these elements so special?

We must realize, following the discussion in Chapters 2 and 3, that life consists of spirit that has taken on physical existence. This spirit arises from cosmic fire or warmth, the essence of which is carried by the lightest of all elements, hydrogen. Sulphur, an earthly counterpart of this fire process, acts to support the infusion of spirit into matter. But living organisms require physical forms or structures, carbon having this primary task.[28] The chemistry of carbon makes it suited for this purpose as it can link with itself or other substances while incorporating inner spaces as it builds life forms. But the forms that are created depend on a life essence to animate them, the carrier of which is oxygen. Meanwhile no living structures or green matter can exist without the activity of nitrogen. Nitrogen is cosmically sensitive and 'knows' how etheric formative forces should connect with physical substance. Consider its position in chlorophyll (Fig. 4.4) where four nitrogen atoms are arranged in proximal positions around the central magnesium. Again, the fact that the atmosphere is mostly nitrogen points to its vital role in the life of all organisms (Chapter 2).

It is difficult to carry thoughts about the mission of these elements without feeling that spiritual powers must lie behind *all* the chemical elements. In ancient times it was recognized that elements derive their primary impulses from the zodiac—carbon, oxygen, nitrogen and hydrogen relating respectively to the four Aristotelian elements: earth, water, air and fire. Chemical elements also have relationships with the planetary spheres, explaining why different metals are traditionally associated with each of the planets and why alchemists were astrologers *par excellence*. The rhythm displayed in the periodic table derives from the twelve zodiacal signs as mediated by seven planetary spheres. As Hauschka wrote, 'Some day we will learn to regard earthly substances as processes come to rest, as the ancients did when they used the term "the ends of the paths of God". Then we will see the specific qualities of a substance as a free force, not necessarily attached to matter, whose place of origin is some distant spot in the universe.'[29]

In plant and animal nutrition, major and minor elements are recognized[30] while many substances occurring in low concentration do not appear to be essential. The above suggests that if all chemical elements connect with cosmic influences, their interplay within the natural environment is analogous to a universe in microcosm. In this sense, even if a substance has no evident direct action it may nevertheless exert a moderating effect on others which do. What is clear is that without cosmic connections elements could have no function in the support of life.[31]

An exploration of alchemical process

Soil minerals are generally regarded as coming from the rocks underneath or transported by water, wind or former ice. Rain is responsible for cleansing the atmosphere of dust from a variety of sources, including volcanic activity and the disintegration of meteorites. Substances deriving from the weathering of minerals then form the basis of plant nutrition. It is claimed, however, that substances can be formed by other means.

Among the challenges of the *Agriculture Course* is the thought that silica and lime may be transmuted, and that substances can be 'rayed into the earth' from the cosmos—silica, mercury, lead and arsenic being referred to specifically.[32] We might wonder whether the cosmic light essence, discussed above, is able to act as the vehicle for such alchemical transformations.[33] Elements in the sun (and contained in the solar wind) emit characteristic wavelengths of light as detected in spectroscopic lines (Fig. 2.6). If energy and matter are interchangeable, then light relating to elements in the sun's corona might conceivably be made manifest in soils through alchemical means. But all this is speculation unless a satisfactory mechanism is found.

The formation and transformation of elements was a subject which exercised Rudolf Hauschka, who, following the nineteenth century researcher Albrecht von Herzeele, saw substance as *owing its origin to life processes*.[34] For Hauschka, element transformations were part of life, even though their quantitative effects are minute for individual organisms. Louis Kervran, too, in his *Biological Transmutations* collected together many instances which show that transmutations of elements can occur within organisms.[35] If we remain in the inorganic realm and purely in the sphere of nuclear physics, enormous amounts of external energy are necessary to achieve element transmutations, normally with the release of harmful radiation. However, it seems that in a living context such transformations are accomplished without obvious or significant energy input. So can such ideas be carried further, or has this subject always to be given the silent treatment and left on the fringe of science?

Transmutations affecting the earth, plants and the human being

The significance of certain elements has already been noted. Before revealing relationships between elements we will consider their relative abundance in the cosmos and on earth (see below).[36] The comparative figures are a source of real fascination and could occupy a volume on their own. The most abundant elements are those whose atomic weights, when rounded to the nearest whole number, are divisible by 4. Such

Relative abundances of selected elements as percentages by weight

Earth's Crust		Meteorites		Human Body		Universe	
O	46.6	O	32.3	H	63.0	H	93.3
Si	27.2	Fe	28.8	O	25.5	He	6.49
Al	8.13	Si	16.3	C	9.5	O	0.063
Fe	5.00	Mg	12.3	N	1.4	C	0.035
Ca	3.63	S	2.12	Ca	0.31	N	0.011
Na	2.83	Ni	1.57	P	0.22	Ne	0.010
K	2.59	Al	1.38	K	0.06	Mg	0.003
Mg	2.09	Ca	1.33	Cl	0.03	Si	0.003
Ti>H>C				S>Na>Mg		Fe	0.0001

elements comprise 86 per cent of the earth's crust. Geochemists see this as evidence of a primordial formation *via* fusion of helium-4 nuclei, yet there are additional pathways if we consider a *living earth* as portrayed by Rudolf Steiner in his description of the stages of our planet's evolution.

According to this, the influence of silica was dominant originally, while later, for the somewhat fluid human being, nourishment permeated us from a protein-containing 'atmosphere'. Originally, rocks were *in a living state*, which gives insight into why oxygen—carrier of the life force—is such a substantial proportion of the earth's crust. In short, the earth in former time was very much more 'alive' than we find it today, and only after separation of the moon from the earth did solid substance take form. This is the background against which we can begin to understand the incompatibility of Steiner's much shorter view of earth history when compared with estimates of geological time based upon radioactive decay series. Let us now consider key element transformations which trace the early history of the earth.[37]

In (1) below, we draw attention to the primacy of silicon and the fact that not only may two nitrogen nuclei be derived from a single silicon but also those of oxygen and carbon—all essential to life. A silica existence gave way to an atmosphere similar to the one we know, although it should not be implied that other substances and gases such as carbon dioxide were not also present. These changes were completed during the Lemurian period, during which the moon's mass was withdrawn.[38]

$$^{28}\mathbf{Si} = 2 \times {}^{14}\mathbf{N}\ \mathbf{(or\ N_2)} \text{ and } {}^{28}\mathbf{Si} = {}^{16}\mathbf{O} + {}^{12}\mathbf{C}$$
$$\text{also } {}^{28}\mathbf{SiO_2} = {}^{14}\mathbf{N_2} + {}^{16}\mathbf{O_2} \quad (1)$$

Superscripts are atomic masses relative to hydrogen, while subscript numerals refer to numbers of atoms. Note that we are not concerned with chemical reactions, merely with transformation or conversion of elements.

While silicon has been described as expressing a universal cosmic sense or consciousness, nitrogen on the earth signifies a sensitivity to the cosmos, an ability to comprehend and interpret cosmic creative forces. This is the significance of our atmospheric nitrogen. The silicon-nitrogen connection is so close there can be little doubt we are involved with a big cosmic sister and a smaller one—a primary impulse, and the ability to listen.

In (2) we consider Steiner's early protein atmosphere, from which he said that lime precipitates—a far cry from the views of a sedimentologist! Yet this is a profoundly important step in the whole of physical world evolution, for the coming together of nitrogen and carbon to produce calcium invites the thought that *cosmic sensing* (N) plus *physical forming* (C) results in a *desire for being* (Ca). A craving for physical existence thus becomes the mission of calcium, or what we may call a calcium process in living organisms. Here, we should refer to comments already made about lime in Chapters 2 and 3.

$$\mathbf{2N\ (as\ {}^{14}N_2) + {}^{12}C = {}^{40}Ca} \text{ or alternatively } \mathbf{{}^{28}Si + {}^{12}C = {}^{40}Ca} \quad (2)$$

In (3) we are concerned with the densification of the earth and its mineralogical differentiation. Iron is approximately one hundred times more abundant in the cosmos than is predicted by a curve expressing elemental abundance against atomic weight. It is clear that iron has been formed via several pathways. Regarding earth conditions, those hinted at here utilize elements already abundant in its earlier history.

$$\mathbf{{}^{28}Si + {}^{28}Si = {}^{56}Fe} \text{ and also } \mathbf{{}^{40}Ca + {}^{16}O = {}^{56}Fe} \quad (3)$$

It is important that ideas on transmutations are not hidden away, as it seems they may take place in the realm of soil and plant nutrition. In his agriculture lectures, Steiner indicates that just as there is a relationship between nitrogen and oxygen in the atmosphere, so is there between 'limestone' and hydrogen among organic processes. He then says that 'limestone and potash are constantly being transmuted into something very like nitrogen, and at length into actual nitrogen. And the nitrogen that is formed in this way is of the greatest benefit to plant growth'.[39] In (4) we have attempted to portray this on the literal assumption that limestone means calcium carbonate.

$$\mathbf{CaCO_3 + {}^{39}K + {}^{1}H = quasi\ N > normal\ N} \quad (4)$$

Remarkably, the nuclear complement on the left-hand side of this equation amounts to exactly ten atoms of nitrogen.[40] On the other hand, this transmutation could be stated in a different form, as in (5), because lime is sometimes a way of referring to calcium alone.

$$\mathbf{{}^{40}Ca + {}^{39}K + 5H\ (from\ H_2O) = 6 \times quasi\ N > 6N\ (or\ 3N_2)} \quad (5)$$

In experiments mainly using plants grown in distilled water, Herzeele demonstrated that their ash content changes as growth proceeds. This led him to conclude that *within living plants* a form of alchemy enables one substance to be transformed into another. The following transmutation series (6) was proposed originating from carbon dioxide and water alone![41]

$$\mathbf{CO_2 > Mg > Ca > P > S} \text{ and in other experiments } \mathbf{N > K} \quad (6)$$

Presenting somewhat similar problems, are people who claim to have lived from little else than light. In order to maintain a healthy organism such individuals must have been able to convert water and trace substances to almost any material required by the body.[42] Nutritional insight into how such processes might occur is offered in Chapter 10. Steiner indicated that in meditation we are able to transmute the lime of our bony substance into nitrogen, resulting in us breathing out more nitrogen than we breathe in! In a sense then, we are able to redeem a token amount of calcium that has been involved in supporting our physical structure.[43] We can formulate (7) to specify the raw materials and product of this particular transmutation.

$$\mathbf{2Ca + 4H\ (as\ in\ 2H_2O) = 6N\ (or\ 3N_2)} \quad (7)$$

A logical basis for element transmutation?

The question is posed in this way because what is or is not worthy of the epithet 'scientific' is for individuals to decide. The problems of establishing the existence of element transmutation are immense because scientific research is based on simplifying questions to a point where relationships can be clearly established. Natural or living systems are normally more complex than experiments can handle, and while life might possibly be monitored, it cannot be analysed. But whatever the current scientific paradigm it is essential for us to freely formulate new questions about our existence as well as revisiting earlier ideas in the light of new knowledge.

In order to bring our discussion of transmutation into sharper focus, we should turn to consider current knowledge of the atom itself. It is generally accepted that the atomic nucleus consists of positively charged particles called protons. But the more curious student might have wondered why adjacent protons did not repel each other and therefore fail to form a cluster. In order for this not to happen, particle physicists specify a force *acting from the centre*![44] This is known as the 'strong interaction' and is a centripetal force which, for each individual element, must exceed the repulsion between clustered nucleons. This force is of profound importance for it makes possible the formation of physical substance *and therefore life as we know it*. We referred above to the enchantment of elemental beings. For what purpose were they enchanted? So that spiritual powers could create the fundamental states of matter on which life depends. Logical thought therefore leads one to conclude that the 'strong interaction' is provided by the existence of elemental beings.

We have discussed the passage of a light ether or essence into the soil and its retention there until utilized by plant life. If we allow that elemental beings—beings also of an etheric nature—lie at the heart of all matter, we are then able to consider that transmutation rests on establishing a resonance between nuclear forces and the essence of living organisms, namely their etheric nature. So the circumstances likely to promote transmutation exist within living organisms or a soil environment permeated by life radiations. As a result atoms, which are largely immutable by physical force, become open to disruption and reformation. In the *Agriculture Course*, Steiner refers to transmutation in connection with soils having a living quality. Such soils would also be more receptive to cosmic influences.

In this connection, earlier alchemists were associated with what was termed the *Philosopher's Stone*. This referred to the 'cosmic formative principle', whether the shaping of physical substances or the formation of thought. Such is the spiritual essence of carbon, which uniquely occurs both as the framework of life and as a precious stone. Reference to alchemists performing transformation of elements into gold undoubtedly overplays one aspect of their life's studies but it would seem that only after creating a *living crucible within the soil or through the medium of the life process* would such transformations of substance have been possible. For achieving their required objectives, Druid alchemists will also have been mindful of timings that offered appropriate formative forces. Besides the time of day, these would have involved moon and planets against a background of the zodiac.[45]

Plant nutrients and the life ether

We must now turn our attention to the upward streaming of substances within the plant, so completing the story which began with discussion of photosynthesis.

What are we to make of plant nutrition? Substances and water certainly are involved, but this upward stream incorporates the life or growing force in plants. And force it really is, for not only does it enable the plant to work against the force of gravity, it is—as I recall as a child—the force that enables a daffodil to push up through an asphalted path!

We need to realize that in daytime and summer as light is received and absorbed by the earth, life forces are simultaneously being drawn in the opposite direction, upwards, by the sun. On an etheric level this is the equivalent of Newton's First Law, namely that every action induces an opposite reaction. These life or growth forces are of course moderated or enhanced by lunar rhythms. Meanwhile, interfacing mineral and organic worlds, the gnomes are key agents in conducting this force, for Steiner describes them as literally 'pushing up the plants'. Their efforts nevertheless merge with those of undines who oversee water processes.[46]

Nutrient processes and a principle of 'offering-up'

It is a matter of simple observation that humans, animals and plants are all dependant on the realms below them for their existence on the earth. So it might equally be said that the mineral realm offers itself up to the plant, the plant to the animal, while minerals, plants and animals all offer themselves in the service of human beings (Fig. 4.13). Indeed, this principle of 'offering up' might well be extended to ourselves, to what we do for our fellow human beings, what we offer to the world in general and to realms spiritually higher than ourselves. There is perhaps an echo here of these words of Goethe: 'Heavenly forces ascend and descend, passing golden chalices from hand to hand.'

But how does this idea apply in plant nutrition? Let us consider substances taken up in ionic form by plant roots. These participate in an 'upward stream' which meets the sunlight in the leaves. From an analytical perspective they are representatives of the mineral world, having physical existence but no apparent life. These cationic or anionic substances, however, all have relationships to life processes, and if soil conditions come near to those described above they will have latent excitation.

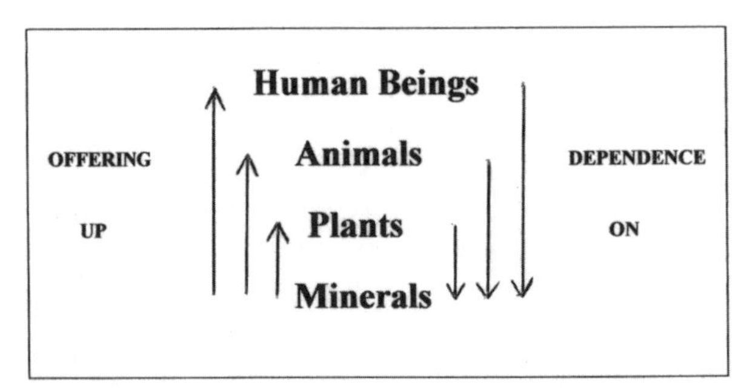

Fig. 4.13 Relationships between the kingdoms of nature

Substance and process in plant nutrition

We should first clarify the relationship between substance and process.[47] Substance is an expression of the physical body of organisms (*bios*). Process is an expression of inner life activity (*zoë*). From the viewpoint of orthodox plant nutrition, substances in solution have individual concentrations. When absorbed, they enter a flow with a concentration gradient, this being a function of metabolic activity. Nutrient ions within the plant are not free to pursue their own passage in a random way but are held according to the operation of metabolic processes. In general, the plant determines the behaviour of substances, not the other way round. The 'offering up' idea slightly transforms this view. Ions of particular substances thus become part of a process within the plant's etheric organism. When they pass over into the plant's etheric body, they release their higher nature. Each becomes enchanted and *rays out its influence*—offering up its own life ether. This is undoubtedly what Rudolf Steiner meant when he said that the oxygen of the atmosphere is 'dead', but becomes 'alive' inside us![48] So instead of visualizing an ion moving on a chemical gradient to the plant tops we should see a radiation spreading out.

Let us now take a further step regarding the transmutation of elements. Earlier we suggested these might arise where a resonance occurs between atomic nuclei and life processes. Here, we must be aware that elemental beings are at work in all plant processes. Gnomes, interfacing the mineral earth, have ultimate wisdom of the nutrient needs of plants. They are effectively sense organs for the plant—a capacity heightened when the biodynamic preparations are used. Their work involves supervision of soil organisms, above all, the mycorrhizae. They literally conduct the orchestra within the soil! The undines, for their part, operate within the watery flow of the plant and are to be regarded as the plant chemists. Instead of always describing processes as if they were inanimate, we should learn to accept that wisdom prevails in the whole conduct of life processes—an invisible workforce is involved. These are undoubtedly difficult ideas and can be too difficult a leap of imagination for many people.

Among the transformations referred to is that involving a 'raying in of substances'. A possible mechanism for this would be via the light ether as we have traced it in this chapter. Such radiations might be utilized by elemental beings to create small quantities of substances judged to be deficient in relation to the needs of plants. Elements that have benefited from contact with clay and humus should also be strongly infused with ether radiation. The conclusion that may be drawn is that nature's ability to form trace amounts of newly formed elements will depend on the level of etheric forces within soil and plant. Having overcome certain threshold energies, such radiations may then be recombined to configure new elements required for healthy growth.

The final step in this overall hypothesis is to propose that nutrient deficiencies will be least likely to arise in situations where etheric radiations are strong. In this connection, organic agriculture systems suffer comparatively little from nutrient deficiencies. Meanwhile, it is periodically asserted by agricultural chemists that such systems give inadequate attention to the restitution of soil micronutrients. Elements such as phosphorus are therefore considered to be depleted by organic farming yet soil and plant analysis rarely support such a view.

Concluding thoughts

In the *Agriculture Course*, Rudolf Steiner indicates that artificial fertilizers make plants relatively dependent on the purely chemical forces dissolved in water. The adoption of hydroponics and use of artificial light will also support a similar kind of growth. On the other hand we have proceeded through this chapter to show how a force other than the purely chemical should be recognized as the basis for a true vitality. As already indicated, this is a force that originates from the constellations and is mediated by the sun in conjunction with the outer planets. This force, conducted by the light of the sun, is received by the earth and nurtured there, *given appropriate soil conditions*. In this way, nutrient substances are subject to a qualitative transformation. There is thus a major difference between a substance originating from a synthetic process and one which has been derived from the cycle of life and which subsequently has experienced an intensification of cosmic influence within the soil. Substance not only has inherent chemical properties but also biography. Among his many experiments, Hauschka showed that while natural substances could be potentized to exhibit rhythmical effects on living systems, synthetic substances had no such capacity.[49]

There is a proverb originating from a work by Alexander Pope that 'fools rush in where angels fear to tread'. Unfortunately, 'fearing to tread' merely perpetuates an impasse and does not move the agenda on. Thus, as a by-product of discussing cosmic energies, this chapter has tackled the vexed question of element transformations. We are drawn to the conclusion that the wisdom underlying nature, as executed by the elemental world, channels appropriate cosmic forces providing a suitable vessel is available. The receptiveness of organic systems to alchemical processes would thus appear as a further, indeed fundamental reason why they are inherently robust and constitute a genuinely *sustainable* form of agriculture.

It is very much the task of biodynamics to enhance the sensitivity of our soils to enable the above processes to take place. Aside from good soil management this is achieved by using a range of preparations which are to be discussed in the next chapter.

5. Supporting and Regulating Natural Processes

We come now to consider measures designed to improve the vitality of earth and crops—the original biodynamic preparations of Rudolf Steiner[1]—together with further ways in which biodynamic, homoeopathic and 'homeodynamic' principles have been applied. Before concluding the chapter, brief mention will also be made of the original biodynamic approaches for tackling weeds, pests and disease.

The lowest common denominator for these various tasks is that of *healing*: healing to restore some of the earth's lost vitality, healing aspects of an agricultural organism which have become unbalanced, and providing protection from environmental stresses.

The biodynamic spray preparations

The biodynamic preparations as originally conceived by Rudolf Steiner, comprise two groups, the field sprays (500 and 501) and those for the compost process (502–507).[2] The word *preparation* refers to an original process by which these materials are created, and as such rather conceals their purpose. In fact, the biodynamic preparations are activators, enhancers, vitalizers or facilitators of processes.

The field sprays regulate the life of plants in a polar opposite way. *Horn manure* activates the soil-root realm while *horn silica* guides the upward development of plant growth. These preparations should be used in combination to bring about balanced growth. But why horns? These were discussed in Chapter 3. We are concerned with their sensitivity to cosmic forces and their capacity to contain those forces accumulated in the manure and silica respectively. We should note that certain Ayurvedic herbal remedies depend on burying material in a horn for a period to strengthen its potency, and traditional doctors overseas still utilize horn caskets for storing their remedies, including snake anti-venom.

As information is available in other sources[3] only essential facts will be given here on how each preparation is made and applied.

Horn manure (500)

For this we require cow horns and cow manure. Fresh cow manure is packed into the horns (the core having been removed) (Fig. 5.1). Horns are then buried in good, well-drained soil, at 15–40 cm depth, with open ends turned downwards (Fig. 5.2). Good topsoil with plenty of compost

Fig. 5.1 Horns filled with cow dung ready for burying

Fig. 5.2 Placement of horns in soil pit

or humus is then placed around and over them rather like the act of planting. This will optimize the working of cosmic energies (Chapter 4). This procedure should take place in autumn, ideally around Michaelmas in the northern hemisphere. This relates to the intensification of the earth's life forces in wintertime (Chapter 3). The location of the buried horns should be clearly marked to avoid embarrassment later! These should be dug up roughly six months later and their contents *either* used as below *or* stored (see p. 120).

A portion of the material from the horn is crumbled into water in a bucket at the rate of 60 g in 35 litres of water per hectare (or 25 g to 15 litres per acre). For large areas, according to supplies of preparation, the amounts can be reduced. Traditionally this is hand-stirred for one hour, making sure to develop a vortex or cone-shaped depression in the liquid (see below). The material is then ready for spraying. As the forces of horn

Fig. 5.3 Alternative ways of applying the field spray preparations. Top left: Bucket method. Top right: Knapsack sprayer. Lower: Tractor or ATV-mounted 3-point linkage kit (Lindsay Ellacott, UK)

manure are to be directed downwards to soil and roots it is logical to apply this preparation in the evening when the earth's flow of energy is in that direction. It should be applied as large drops within 2 hours of stirring, using a bucket and brush (a bundle of leafy twigs suffices), or alternatively by knapsack or tractor-mounted sprayer (Fig. 5.3).

Horn manure is used at the start of any cycle of crop growing, before or immediately after sowing or planting out. It should also be used after a hay cut or whenever grassland begins a new cycle of growth. In polytunnels, it can be used before the start of any new crop. Horn manure connects with processes governed by the inner planets related to germination, root development and vegetative growth as summarized in Chapter 2. It represents a 'highly concentrated, life-giving, manuring force'. The latter is the product not only of the original manure's potency, but of its intensification through infusion with the winter life forces of the earth. Finally, the etheric and astral force of the horn manure needs to be unlocked from physical substance, which is why a particular stirring method is chosen (see below and Chapter 8).

Horn silica (501)

Silica or quartz crystals are illustrated in Fig. 4.1. There are many forms of quartz. Reasonably pure rock silica is good enough for this purpose without searching for perfect crystals! The preparation requires finely crushed silica to be placed into cows' horns. The silica powder is made into a paste with water before filling the horns. The ends of the horns are plugged with clay or soil to prevent subsequent loss of silica. Horns are buried, as for horn manure, except that the period of highest sun should be chosen—from spring to autumn. This corresponds to a period of out-breathing of the earth's life forces (longer day-length), for at this time the light ether, to which silica is strongly related, more actively penetrates the earth (Chapter 4). It is this which the ground-up silica will be absorbing.[4]

Horn silica contains only trace amounts of the original substance and is principally applied for its radiant silica force. It enables the plant to be more sensitive to all those qualities contained in the light of day. It strengthens the form-giving aspect of growth together with quality and nutritive value. Silica enhances the absorption of far-infrared frequencies to which water and organic molecules resonate. This is likely to confer antioxidant properties, which translate into healthier growth and better keeping quality. Horn silica may also be used effectively for the suppression of fungal disease resulting from damp conditions with poor light levels (see also oak bark preparation and *Equisetum* spray).

A 2.5 g portion of horn silica is added to 35 litres of water per hectare (1 g in 15 litres per acre). Stirring is carried out as below. Spray as a fine mist in the early morning, preferably before dew has evaporated. This relates to the earth's daily out-breathing, and enables the silica forces to be carried upwards within the structure of the plants. As the practicality of spraying tall plants may call into question the use of this preparation it is worth realizing that silica forces reaching the lower parts will be transmitted throughout the entire plant.

Crops should have a properly established root system before horn silica is first sprayed. It is then best sprayed at stages of growth where a change of plant morphology is noted. As horn silica can significantly influence aspects of crop quality, repeated application at intervals up to harvest will be beneficial.[5] Applications soon before harvest, when vegetative growth is no longer to be encouraged, can be carried out in the afternoon.

Stirring the horn preparations

The above preparations require to be stirred into a volume of water before spraying. Steiner's original guidance, using materials available to him at the time, was to stir in a bucket to develop a vortex (Chapter 8), doing so in alternate directions, breaking down the vortex at regular intervals to create chaotic motion. Stirring is then recommenced in the opposite direction, the whole operation being continued for one hour (Fig. 5.4).

One should suspect that more happens here than a simple mixing or dissolving process! Water is a special substance and is able to take on the resonance given to it by materials with which it is in contact. The radiant energy from the preparations is thus transmitted to and absorbed by the water. This process should be complete after one hour's duration. It may be objected that there can be infinite variation in the stirring process and

Fig. 5.4 Spiral movement of water into a vortex

that one hour is arguably of little consequence. Indeed, there is some suggestion from dowsing that while one hour gives a satisfactory outcome so also do shorter periods. However, one hour is an exact proportion of the way that time governs our lives and one finds that early apothecaries generally never stirred or ground their remedies for less than this time.

Certain changes tend to take place—increase in dissolved oxygen (and bacterial numbers in the case of 500), together with a slight increase in temperature from ambient conditions, human contact and, more importantly, from frictional energy given to the water. Not surprisingly, it is the general experience that there is a change in viscosity after about 20 minutes, and the water becomes easier to stir!

Depending on the area to be treated, the bucket has become the barrel (Fig. 5.5), while hand stirring has given way in some cases to machine stirring (Fig. 5.6). With ever larger farms turning to biodynamics it is essential that efficient systems are in place to deliver these preparations in a timely manner. All this has naturally led to fierce debate among biodynamic practitioners—firstly the replacement of the human being by a machine, and secondly the precise harm inflicted on the water by its characteristics, in particular that of its electromagnetic field.

A further development has been the use of 'flowforms' (Chapter 8), which engender no less resistance in certain quarters. Issues have arisen over the material the bowls are made of, the type of pump employed or the precise motions of the water as it flows through the system. Flowforms, however, were developed through painstaking research of the natural motions of water together with the shapes produced by formative forces.[6] Besides a general mixing, many combine a lemniscate motion as water swishes across each set of bowls, with micro-vortices as water

Fig. 5.5 Barrel stirring. Left: Traditional wooden cask. Right: Plastic drum and tripod

Fig. 5.6 Stirring machine in New South Wales, Australia

Fig. 5.7 Flowform cascade in North Island, New Zealand

cascades from one level to the next (Fig. 5.7). The same micro-vortices are achieved in the succussion process employed by homoeopaths. Flowforms are the most practical solution for delivering large volumes of spray, particularly where a person has to work single-handed.

While there can be criticism of any large-scale mechanized technique based on assumptions about the energy quality of the resulting water, resistance to machine stirring may simply arise out of concern to defend the original guidance of Rudolf Steiner. A brief evaluation of stirring methods is offered later.

Field spraying—practical considerations

Here, the question of timing needs to be addressed. Current Demeter certification requires producers to apply a *minimum* of one application of each of these sprays in the course of a year. If one chooses to comply literally with this requirement then it is very important that the respective applications are timely. In the case of the 500 where focus is on the land, coverage can often be achieved more or less at the same time and most will attempt to apply this at the beginning of the 'growing season'. But one should avoid having a fixed idea of when this spray should be applied, for autumn-sown crops or following a hay cut also represent the start of a cycle of growth. On the other hand, the 501 is more crop-specific, meaning that there is an appropriate time for treating growing vegetables, field crops and pasture. Horn silica then may not lend itself so well to blanket coverage, except perhaps in all-grassland situations. Because it is to be sprayed in the early morning, many do not make full or effective use of it. And as polytunnels become ever more widespread, we need to be flexible and adaptive in our approach to using both the spray preparations and regard repeated applications as normal practice (see Chapter 6).

Are weather and soil conditions of relevance to the application of the horn preparations? There is often uncertainty here. Does it make sense applying horn manure when the soil is too cold or dry? Can either be applied if rain is forecast—or even falling? Should horn silica be applied during drought conditions? We need to realize that these preparations address life processes, which work through the medium of water. Horn manure activates the plant root, its relation to the clay-humus and soil micro-organisms. If soil moisture is scarce there is little, apart from root tissues themselves, that is receptive and therefore able to capture the essence of horn manure. More obviously, little growth is possible. Equally, little benefit can be expected if temperatures are too low for soil biological response. In cold weather, the practice of warming water prior to stirring horn manure may well help to release ether forces but application of the preparation at such times would seem misguided.

In a climate where it is often raining or forecast to do so, one worries about the efficacy of sprays. To apply 501 when steady rain is falling would be absurd, but we are dealing here with radiant energies which, once in contact with plant surfaces, will disperse themselves within the biological tissues and connect to the 'being' of the plant. To think otherwise is to adopt the mindset of someone applying pesticide. For this reason application of 501 to the lower foliage of taller plants will suffice, instead of endangering life and limb with the use of steps and ladders. Similarly, we should expect horn manure to penetrate the ground without having to physically infiltrate. However, if the ground is waterlogged the soil is not able to support life processes, so it makes no sense applying 500 at such a time.

On the other hand, if horn silica is applied when plants are under moisture stress there is no support for vegetative growth; consequently there is a risk of premature ripening and reduction in yield. This is a factor which often restricts choice of 501 spray timing within the spring and summer season. But as horn silica assists formative growth and ripening, it may be of supreme value under prolonged dull, moist weather conditions, when it can contribute those light forces which were concentrated in the silica horn. Winter conditions are frequently mild so pasture grass continues to grow and animals can be kept outside provided poaching is not a problem. But these days are often cloudy, damp and gloomy. Is this not an opportunity to use the silica preparation? And if it is considered that temperatures are limiting, then why not combine it with valerian as below. Furthermore, given an inventory of newer preparations (outlined later), for successful growing one should not remain over-reliant on the original sprays.

We should realize that the biodynamic preparations, through their impact on soil and plants, support animal life too. Animals will thus benefit directly from grazing land that has been treated. Often forgotten is the free-range hen which is out all year. In view of recent disease outbreaks we should be redoubling our efforts to strengthen the bird organism.[7]

The biodynamic compost preparations

Compost preparations moderate the decomposition process and promote the formation of stable humus. They direct the process of plant nutrition through helping plants achieve a balanced use of the different chemical elements. Consideration of the processes they address reveals connections with the various planets.[8]

There are six of these preparations of herbal origin, four of which are enclosed within an animal organ.[9] While these may resemble the original herb material, each has effectively been transubstantiated—it not only has a *life* quality but also a capacity to *sense* the needs of plants.[10]

Each individual compost preparation is located separately within the compost either when heaps are half made or at the end, before covering (Fig. 5.8). The reason for this is that their individual influences may better express themselves throughout the mass than if completely mixed. A mere

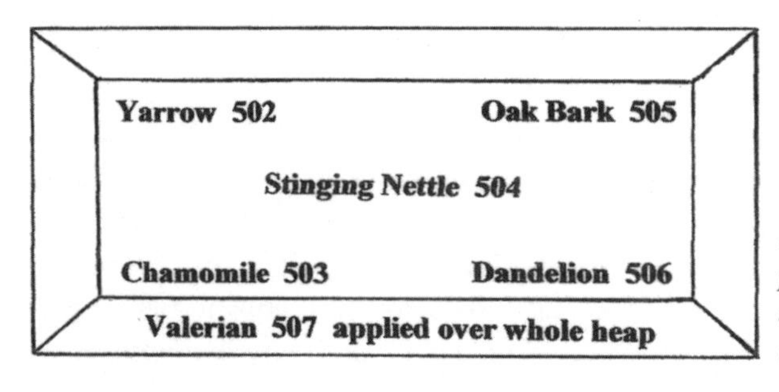

Fig. 5.8 Normal pattern for placing compost preparations in a compost heap

fingertip amount of each solid preparation is sufficient, often introduced into a ball of soil and dropped into individual holes. Diluted juice of valerian is then sprinkled over the entire heap. Compost preparations are collectively referred to as a 'set'. This consists of 1 cm^3 each of the solids plus 2 cm^3 of valerian, and is considered adequate for treating up to 8 m^3 of compost.

The compost preparations can also be added to liquid manures and slurry. The author has witnessed three different approaches to this. In one, the individual preparations are enclosed in soil or compost and thrown into the barrel. In the second, they are suspended 10 cm under the liquid surface from floating wooden laths tied crosswise. In the third, they are placed in nests of fibrous material which float independently on the surface. In each case, diluted valerian is added. Nothing like choice! But one ought to be reminded that compost, to which we more commonly add these preparations, should be aerobic, so the liquid organic fertilizer must have access to air movement and be periodically stirred to avoid foul conditions lower down.

A description and interpretation of the role of each of the compost preparations now follows.

Yarrow—Achillea millefolium *(502)*

This is based on the flowers, which are picked in the early morning when they are in full bloom. They are then hung up in the dry during summer. The dried yarrow blossoms are then packed into the bladder of a stag and buried in the ground for the winter half-year. This can be placed between clay pots at a depth of about 20 cm and surrounded with loose topsoil.

Yarrow manifests for the plant kingdom a perfect quantitative relationship between potassium and sulphur, between earthly and cosmic influences on plant growth. Potassium is representative of earthly salts in the sense of the earlier alchemists while sulphur expresses the working of cosmic archetypal forces into the shaping of living matter. No spiritual forces can work in the organic world without sulphur and its crucial role in protein formation. The element relationships in yarrow are said to promote an ideal patterning for this. So the yarrow preparation is concerned overall with the sensitivity of nutrient substances to cosmic forces. With a strong connection to nutrition and with the bladder's relation to its counterpart, excretion, we see the influence of Venus.

Chamomile—Matricaria recutita *(503)*

This is from the flowers of the herb which are again picked early in the morning. Only the flower heads should be taken. These are spread out and dried in the shade. They are then placed in the small intestine of a cow, making a number of sausage-like sleeves about 10 cm long. These are then buried over winter in the soil in a protective box which can be packed with peat.

Chamomile's particular relationship to sulphur causes it to 'attract' calcium in addition to potassium. Calcium supports health through linking the life force with the carbon structures of the physical organism (cell communication). In this respect it has a kinship with nitrogen activity. For these reasons the chamomile preparation engenders nitrogen stability within the compost. Intestines are connected with movement processes and, as such, with the activity of Mercury. The 'caduceus'—a symbol of healing—is associated with Hermes or Mercury. Chamomile is well-known for bringing about a healing effect when the metabolic system has become disturbed.

Stinging nettle—Urtica dioica *(504)*

The whole herb is taken before it comes into flower. Again, plants are gathered in the early morning. Dig a soil pit, line it with wood, place nettles inside, then put a lid on top and cover with soil. Further nettles can be added later according to availability, but then the full complement of nettles should be covered with soil and left for a full year. No animal organ is required in this case.

Here we progress from processes connected with the alkali and alkaline earth elements to the metals in the middle section of the periodic table. Of major importance here is iron, traditionally connected with the planet Mars. The nettle is said to regulate those processes connected with this element. The nettle preparation influences compost by controlling excessive rates of decomposition. Steiner refers to the nettle as a 'regular Jack-of-all-trades who can do very, very much'. When talking of the human bloodstream, Steiner connected the force of individuality with iron radiations. This preparation therefore appears to confer on compost and soil a 'sensing' for the *individual* needs of crops. This property, allied to consciousness, is an attribute of outer planetary influence.

Oak bark—Quercus robur *or* Q. petraea *(505)*

The oak bark should be taken from the outer or dead woody layer of a mature tree. When broken up into very small crumbs it is then placed inside the skull of a domestic animal: a cow, sheep, goat or horse. The bark should be moistened and put into the hole where the spinal cord begins, then sealed with a piece of the animal's own bone or a piece of wood. The skull should then be placed inside a protective case and buried in a situation where water is likely to flow past it for a time during the winter period. This can be achieved by a river bank or spring, or where water is likely to be soaking away around buildings.

This preparation regulates excess moon forces which in wet spells have a tendency to cause plant diseases. When life forces are too active, organisms normally associated with decomposition gain access to this surplus energy and begin to parasitize the plant (see also Chapter 3). Oak bark contains calcium, which acts to hold in etheric or life energies. For it to have a healing effect the lime must be taken from where it has originated through life processes. Here in the bark of the oak tree and well above the soil surface, it is in a uniquely living form. The bark calcium is then placed into that organ which has been connected with the animal's consciousness. Within the skull, the bark experiences the earth's gathered winter life force under circumstances of excess water, implicitly connecting it with lunar influence. This experience engenders a 'beneficial contemplative influence' which opposes the excessive regenerative force so threatening to plant health. In this way, oak bark preparation is prophylactic against plant diseases.

Dandelion—Taraxacum officinale *(506)*

The flowers of this herb should again be picked early in the morning when they are young and before the centres of the florets are fully opened. They are then laid out to dry for 2–3 days and covered to inhibit the tendency towards seeding. The wilted flowers are then packed into sewn-up bags made from the mesentery of a cow or ox. (The mesentery is the lining of the digestive cavity sometimes referred to as the cole fat.) The individual bags should be about 10 cm in diameter. They are then hung up in the shade for a month inside a protective net. After this, they are stored in a peat-lined box until autumn when, with suitable protection from animals, they are buried in the soil until spring.

The dandelion is described as a 'messenger from heaven' as its role is to draw in a cosmic silica force, silica being present homoeopathically in the atmosphere. This enables plants to be sensitive to everything in their surroundings. To accomplish this task a correct relationship between 'cosmic' silica (symbolic of the light ether and a connection with Jupiter) and 'earthly' potassium must be brought about. Such is the nature of dandelion that this will be achieved through experiencing summer's influence above ground and the earth's concentrated life forces while buried in winter.

The significance of the mesentery is that it is a membrane which encases and protects major internal organs which do not experience outer sensations. Such membranes are sensitive and able to store images, and therefore memories. By enclosing in the mesentery, the dandelion substance effectively becomes an organ, retaining outer impressions drawn in by the silica. The preparation therefore enables plants to be highly sensitive to their surroundings and therefore to the organism of the whole farm. A kind of alchemy is involved, for plants 'then benefit not only by what is in the tilled field, but also by that which is in the soil of the adjacent meadow, or of the neighbouring wood or forest'.[11]

Valerian—Valeriana officinalis *(507)*

The flowers of this herb should be picked in the evening when fully out. These have to be minced, then wrapped in cloth and pressed to extract the juice. The latter should be stored in dark bottles with airtight closures. No further treatment is necessary. Diluting and stirring is all that is required. Take 2–3 cm^3, or one teaspoonful, of valerian juice and dilute with about 5 litres of water, preferably rainwater. Stir or shake for about 10 minutes before applying as large droplets to the entire exterior of the compost heap. In the case of a liquid fertilizer brew, it is better to add 100–200 ml of the diluted valerian and stir.

Valerian, with its pale blue flowers, harnesses the distant forces of Saturn, closest of the traditional planets to the 'fixed' stars—the regions of the plant archetypes or formative influences of the plant world. The association with Saturn indicates that we deal here with the working of the finest of all ethers, that of warmth, traditionally connected with phosphorus. By sprinkling over the compost, the heap is enclosed by a kind of warmth sheath. This is subsequently of benefit to plant growth, for if the 'warmth organism' of plants is not sufficiently maintained their connection with archetypal forces is weakened. In alternative medicine, phosphorus has an ego-strengthening effect. In supporting the phosphorus metabolism of plants this preparation therefore strengthens the connection with their own ego or essential formative forces.

Storage of preparations

In general, biodynamic preparations should be stored in a cool, dark place. Use a box filled with moistened peat, as this is the very best material for insulation from extraneous radiations. The box should have a lid, preferably peat-filled.[12] It is common practice to place each preparation in its own earthenware pot with a lid. By contrast, horn silica is best kept in a sunny place—usually on a window sill.

Storage should always be a temporary measure for the preparations, as radiating substances will lose their activity over a period of time. They should be regarded as of extremely doubtful potency after more than two years of manufacture. In any case the object of biodynamics is not to make a museum out of the preparations store but to actively deploy preparations on the land!

The human factor in the use of the preparations

As with all methods, rather than simply following instructions, it helps to develop understanding of the actions you are taking, for the human being really does carry a creative force. This leads to better motivation to carry out the various operations, particularly if time presses. It should also improve the effects of actions which are essentially human gestures towards the earth and its life processes. The basic thought to have in mind concerns the earth's breathing rhythms and the elemental forces which are connected with it. Because seasonal rhythms are disturbed or being lost, the preparations have an increasingly important role in supporting elemental beings.

Science insists that for a technique to be valid the human being should have no effect on the outcome. Anyone doing the work could therefore be expected to achieve the same results—that is repeatability. But for working with living things the human being does interact with the life process and it therefore *will* matter that a sympathetic attitude is displayed, for example by those involved in stirring and applying the preparations. Here, as in other actions, it is necessary to live in the moment rather than with one's mind on what has to be done next!

There are many separate operations involved in making and applying biodynamic preparations, so not surprisingly views are held about exactly how this or that operation should be conducted. While there are essential steps to carry out, exploration of previous biodynamic literature suggests that many points of detail are purely matters of opinion or preference. That is human freedom. The only danger is that newcomers may easily adopt certain 'shibboleths', not knowing whether they are really essential! The majority of those using biodynamic preparations purchase them from elsewhere, so there is a lack of involvement with or knowledge about the circumstances of their manufacture. It is for this reason that the stirring method attracts so much attention.

However limited one's understanding of biodynamic measures Steiner assured his audience that they will still take effect, and we are encouraged to apply the preparations as widely as possible. But in addition to what the preparations accomplish of themselves, their use engenders in us a special attitude of mind towards the living earth.[13] Biodynamic practices are, however, by no means the only path by which human beings draw close to nature. Particularly in the past, a natural reverence pervaded rural society, while more recently gestures of individual 'reverence' have taken different forms, including the use of bird song or classical music played from a tractor.[14] Formerly it was said that *the best manure was the farmer's boot*, and in China a similar saying referred to *the shadow of the farmer.* Steiner emphasized time and again the importance of each individual forming a special relationship to what they do. In this case, aspects of biodynamic practice which could be considered imperfect may well be compensated for by a conscious identification with the sacrament being undertaken. Whether in an age of certification Steiner's ideal becomes more or less difficult to achieve is for individuals to decide.

At this point it seems appropriate to consider how various individuals have chosen to work with, and to develop, biodynamic ideas.

Further biodynamic and related preparations

It should be realized that in his agriculture lectures Steiner was laying foundations for a new agriculture rather than presenting a finished product with validity for all time. In Chapter 3 it was explained why cow manure is of such importance in biodynamic land work. Not surprisingly, various developments have occurred, aimed at concentrating its effects by composting with repeated doses of the compost preparations. Manure concentrates were inspired by Max-Karl Schwartz and Nicolaus Remer who used birch-lined pits (Birkengrube). However, the best known and most widely acclaimed development has been the 'cow pat pit' or 'barrel preparation' of Maria Thun.

Cow pat pit

This composite material (CPP) is a useful vehicle for spreading the influence of cow manure and the compost preparations on land not normally receiving compost or which is not normally grazed by cattle.[15] It is very suitable for land in conversion from conventional to biodynamic management and even for addition to compost which has not received compost preparations. It can be incorporated into horn manure, into liquid fertilizers and is also suitable for seed priming (Chapter 7).

To make it, choose a site with good drainage. Dig a pit 60 cm by 1 metre, and 50 cm deep. Make a bottomless box of untreated timber or simply a wood lining with supporting stakes. These dimensions, while approximate, provide a basic working unit which can be added to according to the quantities required. The upper rim should stand a little above the level of the surrounding ground. The best and longest-lasting construction is made of builder's bricks (Fig. 5.9).

Five bucketfuls of fresh, fairly firm cow dung are required. If the cow manure is like slurry it is unsuitable, but if a little on the liquid side then it needs to be well aerated and mixed with a little straw. Add at least 100 g of finely crushed eggshells and 500 g ground basalt—a volcanic rock dust.

Fig. 5.9 Brick-lined cow pat pits under cover in Sri Lanka

Mix very well to create a dynamic wholeness. Fill the excavation to a maximum depth of 25 cm. Insert *three sets* of compost preparations (502–506) by pressing them into the manure mixture to a depth of 5 cm. Stir *three portions* of Valerian (507) for 5 minutes in 350 ml rainwater, then sprinkle over the manure. Cover with moistened hessian sacks so that moisture is retained. Unless the whole operation is under cover, provide with a waterproof lid raised at one side to allow water to run off and air to circulate.

After 1–2 weeks, aerate the manure with a fork (moister mixtures will require earlier and more regular aeration). Add water if necessary but leave the surface flat to avoid excessive drying. The cow pat manure mix will be completed in 2–6 months according to ambient conditions. Cow pat pit is prepared for application as for horn manure (500), but 2–3 times the quantity is normally used, and it is usually stirred for just 20 minutes. If incorporated into horn manure, it is added for the last 20 minutes of stirring, and at the rate of about 50 g to 5 litres (50–60 litres being needed per hectare).

A particular version of this preparation, developed by Christian von Wistinghausen, is known as Mäusdorf compost activator.[16] This can be sprinkled dry or watered between compost layers—equally it can be watered over manure or crop residues to assist balanced decomposition.

Podolinsky's 'prepared 500'

A further initiative has been that of Alex Podolinsky in Australia. He has created a 'prepared 500' by combining the compost preparations with the cow manure before filling the cow horns and burying them. While clearly having different *qualities*, the latter has a similar application to CPP. It is an attempt to address the impossibility of applying biodynamic compost over wide areas, particularly in the case of Australian farms.[17]

Hugo Erbe's preparations

Erbe's well-known 'Three Kings' preparation recognizes the Christian festival of Epiphany (6 January) and, like the other preparations he created, is a sacrament or offering to ensure the continued support of helpful elemental beings. There is in this preparation a real sense that spraying is intended to protect the farm entity within.[18] It is just such a concern over the allegiance of helpful elementals that inspired Steiner's preparations, bequeathed as they were out of a concern for the earth's diminishing life forces. This argument is no doubt reinforced by the effect of urban, industrial and technological developments in recent times. Hugo Erbe's creative repertoire included some 21 preparations which involve varying degrees of complexity to make. These include carbon to improve the earth's breathing process, a preparation to strengthen fruit trees, another to protect them from frost at blossom time and yet another to protect against wild animals.[19] Such ideas have now been carried forward in a more user-friendly, homoeopathic way (see below). Erbe also devised a clay preparation for enhancing the forces 'which mediate between calcium and silica processes' in the soil.

Horn clay

In the previous chapter we examined clay in some detail. More recently there has been interest in the potential of an intensified clay preparation although this had been anticipated by Steiner.[20] As with Erbe's idea, such a preparation is intended to raise the capacity of the soil to retain and transmit those forces essential to healthy plant growth. Its most noticeable benefits will be where alumino-silicate clays are scarce. Typically, this will apply to sandy soils and throughout the tropics and Australia.[21] In the past, marling used to take place in Britain for the remedying of sandy soils—today this is impractical. Even adding clay to compost—as bentonite for example—is costly and without prospect of results unless maintained over many years.

Starting preferably with a high CEC clay, this preparation can be afforded different qualities by burying the clay horn either in summer or winter. It might also be enhanced by addition of basalt rock dust or, if available, the pink potassium feldspar, orthoclase. Where the soil has low clay content a winter-buried horn clay would be suitable, whereas if the soil is very heavy—and often too wet—a summer impulse towards light and warmth might be beneficial. Complementary effects may also be obtained according to whether one is growing roots or seed crops. Using the same basic technique as for horn manure or horn silica, horn clay should be considered part of the biodynamic tool kit and must now be more widely tested.

Therapy for trees

Given the current range of preparations it is possible for biodynamic practitioners to work quite logically towards solutions for many farming and gardening problems. Take the example of trees. Many trees at this time are struggling to maintain their health—they manifest various diseases, deformed growth and pest attack. For a tree, a conventional soil treatment is unrealistic while treatment of external symptoms is an ongoing and often fruitless pursuit. The method described below makes use of the biodynamic principle that the cambial layer of a tree trunk is effectively soil.[22] The aim is to give a tree the capacity to heal itself. An approach which has proven highly effective is the use of a 'tree paste'.

For this, the original recipe consisted of 1 part dried blood, 2 parts diatomite, 3 parts clay and 4 parts cow dung.[23] Neither the precise ingredients nor the same proportions appear to matter, so in practice take portions of silica sand (or diatomaceous earth if available), clay (potting clay or bentonite can be used) and a rather larger volume of cow dung, then mix together to the consistency of a thick slurry with some stirred horn manure. Apply this by brush to the base of the trunk as if you were painting a wall, except you will need to rub it in well. Take it up to a height of at least one metre. It will probably be most readily absorbed into the cambial stream of the tree if the outer bark is moist. In some instances abrasive treatment might help but care should be taken not to harm the tree in this way. While giving thought to seasonal flow of sap, it would be worthwhile observing the calendar timing most appropriate for the use required of the tree (Chapter 6).

Ready-potentized preparations

As we have seen, there has been exploration of biodynamic principles beyond the original *Agriculture Course*. Just as Maria Thun demonstrated the use of homoeopathy in conjunction with insect and weed 'peppers' (below), potentized preparations are now available. Starting in the late 1970s, Glen Atkinson and Enzo Nastati, together with Greg Willis and Hugh Lovel, have used various preparations in potentized form to address the concerns of farmers and growers, the theory behind these being largely in the public domain.[24] Atkinson's preparations have been adopted by conventional growers while the efficacy of a number of his preparations has been demonstrated in field trials carried out by New Zealand's HortResearch. The scope of his products covers the biodynamic preparations and includes the enhancement of photosynthesis, and protection against frost, birds and wild animals.[25]

Meanwhile, in Italy, Enzo Nastati has long been concerned that the earth's environment has so deteriorated since Steiner's time that the existing preparations need to be supplemented and their range extended. Initially set up to regenerate seed quality, over 100 different 'homeodynamic' preparations are now available, covering crop nutrition, pest control and protection against climatic and pollution hazards. As with Podolinsky's 'prepared 500', a number of these result from combining the forces of existing biodynamic preparations. In this respect they represent an evolution of Steiner's original preparations, and their effectiveness is again increasingly supported by field trials.[26]

How can we compare these materials with the 'traditional' biodynamic measures? Certainly there are detailed differences in the way that human beings participate in the process, but it could be argued there is no fundamental difference between the way that traditional preparations are delivered to land and crops and the use of preparations potentized in the manner of homoeopathic pharmacies. In both cases we deal not with substance but with radiating energies, and in either case the response to the earth's 'breathing' process should be the same. If we consider the traditional preparations, these are materials carrying etheric and astral principles (life forces and sensitivity). Before application to the land, however, the necessary volume of water has to undergo an activation process so that it can then convey this force. The difference with a potentized preparation is that the force is already concentrated in a liquid (tincture) which is then simply diluted in water at the time of spray application.

Given the multiplicity of current agricultural problems, potentized preparations offer the producer a flexible and time-efficient holistic approach. Besides their diverse applications, whenever the original preparation materials are scarce or difficult to obtain, potentized preparations offer a solution, together with improved scope for achieving Steiner's vision of covering the widest possible areas.[27] However, while such preparations are 'permitted for use in organic and biodynamic agriculture', they are currently not permitted as substitutes for the original biodynamic preparations for Demeter certification.

The use of radionics

Radionics is described as a method for sending precisely defined healing energy to people, animals or plants wherever they happen to be. It is an extension of the principle of healing by the laying-on of someone's hands. Radionics practitioners are therefore either proven healers or purport to be, as a result of using their equipment. All healthy living organisms are characterized by a particular energy frequency, what we have called here etheric or life energy. In the case of a sick person these energies exhibit abnormal characteristics, so it is the task of the healer to correct them by applying a healing frequency.

A major purpose of biodynamic preparations is to effect a healing and harmonizing process for the earth. Radionic devices—known as field broadcasters or cosmic pipes—are occasionally used to transmit the effects of biodynamic preparations to surrounding land, possibly also using specimens of particular plant pests to actively discourage their prevalence (Fig. 5.10). In this operation it is assumed that the influence of the

Fig. 5.10 Radionic device, North Island, New Zealand

preparations is continuously transmitted while the device is fully connected. But in the absence of spraying or field-walking, many in the biodynamic movement are opposed to this approach. *In extremis*, this view claims that radionics works with forces of sub-nature rather than those which are 'super-physical' and connected with the realm of life forces.[28] However, in view of earlier comments, I would simply say that it expresses a personal relationship to these matters on the part of the operator, who, no less than those applying preparations in the normal way, should be in a position to evaluate its effects—either immediately or in the course of time.

Evidence of the working of the Steiner preparations

Readers will wonder what effects the application of spray and compost preparations have on farm production and the environment. The immediate answer is that both quantitative and qualitative effects are observed. Having myself presented results of experiments to different audiences, it is interesting that people divide themselves into two camps. There are those who insist that in order to justify the biodynamic method there must be measurable results, while others quickly lose interest when detailing different experiments which show these effects. They are quite content to accept that in a field where we deal with invisible forces we should not necessarily expect to see measurable results. While such an attitude may arise from conviction about the validity of biodynamics, it is not logical, for there is no reason to suppose that invisible energies would not give rise to characteristics that were either quantifiable or in some other way perceptual. Furthermore, as we shall see with the testing of food quality in Chapter 10, there are qualitative as well as quantitative indicators.

Evidence for the effects of the biodynamic way of working abounds in anecdotal form. In countless instances, farmers and gardeners themselves perceive differences in the quality of the land and crops, and some also experience the activity of elemental beings, especially at or after the application of preparations. Reports indicate the responses of wildlife on land treated with the spray preparations, and of cattle concentrated on an area of conventional pasture treated with horn manure for the benefit of a sceptic neighbour.

Experimental evidence for the working of the preparations is of many kinds. Pfeiffer,[29] in some of the earliest work, conducted tests on maize and legume seedlings in nutrient solution to discover the root and shoot responses to addition of combinations of the preparations. These showed that when complete sets of the preparations were used there was an advantage over controls, although statistical analysis was not carried out. Perhaps the most remarkable single discovery was that when a standard nutrient solution was made up without lime, the addition of horn manure, chamomile and oak bark preparations compensated for this, as compared with controls. Pfeiffer was also first to show the potential advantage of exposing seeds to very dilute solutions of the biodynamic preparations—seed baths (see biodynamic seed treatment in Chapter 7). He also famously demonstrated, in a box experiment, the preference of earthworms for biodynamically treated soil!

Experiments at different sites have shown that the compost preparation set appears to control excessive heating in the compost heap, while extending the heating period. Ingo Hagel showed that compost preparations sealed into glass vials elicited particular growth responses as

Experimental treatment of spring wheat (var. Jubilar) with horn manure (500) and horn silica (501) with all plots receiving composted manure
+ indicates statistical significance (after Speiss, 1979)

Year	Control	3 × 500	3 × 500 3 × 501	6 × 500 3 × 501
1973	100 = 3.0 t/hs	109	117+	121+
1974	100 = 4.14 t/ha	106	109	111+
1975	100 = 4.1 t/ha	105	102	102
1976	100 = 4.2 t/ha	105+	104	109++

compared to controls, thus demonstrating the radiative principles on which the preparations are based.[30] Although compost is always variable in its nitrogen content, one study found that biodynamic compost contained 65% more nitrate than the control (due to greater stability of nitrogen), as well as significant differences in microbial characteristics.[31]

An important body of biodynamic research has been carried out at the biodynamic research centre (Forschungsring) at Darmstadt, Germany on replicated field plots. This has always extended over a number of seasons to reflect a normal rotation cycle.[32] Thus, over a four-year period, Abele showed raised crop yields resulting from full biodynamic treatment, as well as higher residual soil organic carbon levels. Speiss demonstrated the effectiveness of biodynamics over standard organic practice in his experiments with wheat, sugar beet and carrots. In the wheat experiment, he showed that repeated applications of horn manure and horn silica resulted in increased yield, which for the most intensively treated plots amounted to a 10–20% gain over controls (see table above). Similarly, E. von Wistinghausen showed that when spinach, bush beans, potatoes, carrots and rye grass were grown with fresh or composted manure and combinations of the preparations the best results, consistently exceeding controls by 5%, were obtained when composted manure was treated with the full set of preparations.

So in the various field experiments reported, one or more biodynamic comparisons have been shown to give the best performance, with the distinct indication that repeated application of the field sprays has a beneficial impact. Part of the reason for these favourable results may well be explained in an earlier study by Ahrens, shown below. Here, in an experiment involving micro-organism culture, we see the balancing and harmonizing role of the preparations, notably in achieving higher levels of fungal organisms.

The Darmstadt experiments are on light, sandy soils of inherently low fertility where small additions of organic matter can more easily improve growth performance. In fact, Joachim Raup and Uli König have observed

Numbers of micro-organisms cultured from 1 g dried samples of soil and compost using the Koch plate method (after Ahrens, 1962)

Micro-organisms	Conventional soil	Biodynamic soil	Compost	Compost + BD preps	Culture medium
Bacteria	25.3 M	19.75 M	590 M	750 M	Standard II
Azotobacter	168	506	n.d.	n.d.	Mannit agar
Fungi	108,433	160,493	900,000	2.0 M	Malt agar
Streptomyces	192,771	135,802	3.7 M	3.1 M	Oat agar

a pattern in the outcome of applying the preparations to different types of land. While biodynamics scores well on poorer soils it gives little or no quantitative advantage on more fertile land.[33] This suggests that biodynamic measures moderate any tendency for lush growth and related pest problems.[34] By contrast, they provide more support for plants where conditions are challenging. In this respect, as soil nutrient levels decline, the role of soil fungi becomes more important. Following the Ahrens experiment, it seems that the preparations may be particularly effective in stimulating soil mycorrhizae. This would help explain the success of biodynamics in Australia and New Zealand on poor or degraded soils, as well as more generally in tropical and subtropical areas.

A major study in New Zealand showed that biodynamic farms in general had higher soil organic matter, thicker topsoil, more earthworms, more diverse microbiology, better soil structure and lower soil bulk density than organic or conventional farms.[35] A further comparative study of 12 biodynamic and 12 conventional dairy farms in Australia showed that under equivalent stocking the biodynamic pasture on average required irrigation only every 15 days, as against every 7 days for the conventional farms.[36] Long-term experiments by Pettersson and von Wistinghausen together with more recent work by Mäder and colleagues clearly demonstrate the benefits for soils and root systems under biodynamic management.[37] Bachinger, at Darmstadt, conducted detailed studies of root ramification on soil cores from plots managed according to conventional, organic and biodynamic practices. These showed consistently higher root densities at 5, 15 and 25 cm soil depth for the biodynamic plots[38] (Fig. 5.11). There is, then, convincing evidence on the basis of long-term experiments that biodynamic management leads to the improvement of soil characteristics. A prime contributor to this success must surely be the use of horn manure.

The question is often asked whether different stirring methods render the preparations more or less effective. Research based on replicated plots with wheat carried out at Emerson College, UK by Freya Schikorr[39] showed firstly that stirring gave a significant yield advantage over those treated with unstirred preparation, and that hand stirring produced a better result than automated machine stirring and flowform methods. Recent work also concludes that hand stirring is superior to machine stirring but a full evaluation which includes different types of flowform is still awaited.[40]

The scope of this chapter unfortunately does not allow justice to be done to the increasing amount of high quality organic and biodynamic research being carried out and to the considerable collaboration which is

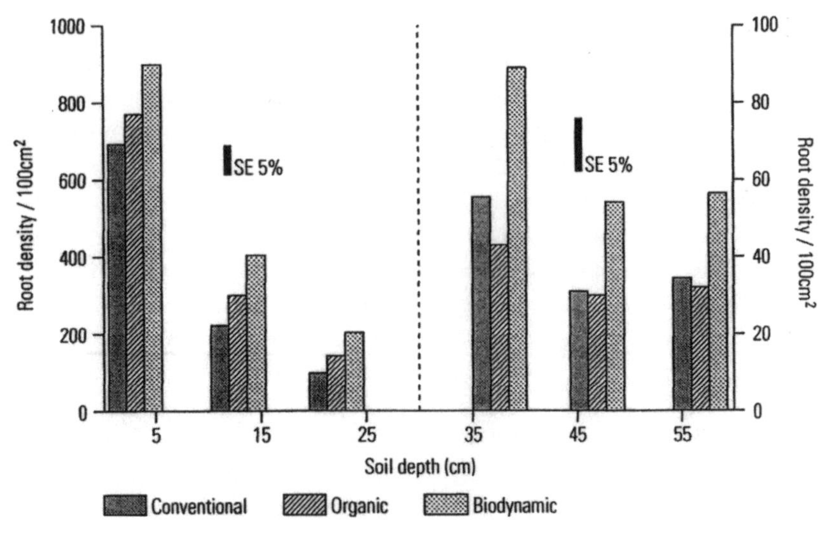

Fig. 5.11 Root density under different management systems (after Bachinger, 1992)

taking place particularly between Institutes in Germany, the Netherlands, Switzerland and Austria.[41]

Weeds, pests and disease

If one is to foster biodiversity as well as support nature's invisible forces, it makes little sense adopting a hostile approach towards organisms that interfere with land work. Although the environment as a whole may be in a state of flux, plant disease and pest problems can also reflect an unbalanced or failed farm organism. The first stage is to establish whether a particular problem is peculiar to the current year or is perennial. For example, weeds in a horticultural operation are usually more of a problem in a wet year, but if they are perpetually a serious issue then one needs to be critical about the way the land is currently managed and be aware of wider issues besides.

Here, the biodynamic approach to weed and pest problems outlined by Rudolf Steiner will be briefly mentioned.[42] In the *Agriculture Course*, Steiner particularly notes the influence of the moon on fertility and how strongly this works through the element of water.[43] Living organisms, including the seeds of plants, encapsulate this lunar force. If this life energy is passed through a fire process—the opposite pole to water—a negative and suppressive influence is brought about. Maria Thun has found that fermentation, also a destructive process, can have a similar effect.[44]

Burning or fermentation of weed seeds

These methods involve the destruction of the seeds of unwanted plants and the application of the processed material to the land to inhibit subsequent growth. One method involves burning seeds, then *either* scattering the ashes *or* making a homoeopathic potentization to the eighth decimal (D) potency. The alternative method involves fermentation of the whole plant with its seeds and then applying this material as a liquid spray. With this method, the nutrients extracted by the weed species can be returned to the land, for the presence of weeds can be an indication of nutrient deficiencies or other soil limitations. The ashing method will be described.

Weed seeds are collected, preferably around full moon, and burnt in a hot fire to ensure they are properly ashed. A metal or other suitable bowl is used for this purpose. Take the ash and grind in a mortar with alternate clockwise and anticlockwise motions (known as dynamization). The dry powder, known as a weed pepper, is applied to the land early in the growing season and will usually need to be mixed with sand or more wood ashes to increase the volume for scattering.

Alternatively the ash can be potentized, a process of successive dilution and shaking used in homoeopathy. Add 1 part of ash to 9 parts of water (by volume). Shake rhythmically in a closed vessel for 3 minutes to obtain the D1 potency (a process known as succussion). Systematically continue this process until D8 is obtained, building up volumes as you progress in order to have adequate D8 for spray application. Spray D8 as for biodynamic 500, using 10–40 litres per acre depending on spraying equipment. Larger volumes can more easily be stirred, creating a vortex, or shaken by using a tipping barrel arranged on a pivot.

The method is best described as one of regulation rather than eradication. While such treatments usually require to be repeated, perhaps for a second or third season, it is reported from New Zealand, for example, that one drop of dynamized seed ash made to D8, then one drop of D8 in 500 litres of water has suppressed nodding thistles within one season! Others, it is fair to say, have not had such immediate rewards.

Insect and mammal pests

For insects, burn a quantity of the particular pest or pests when the moon is in front of the Scorpion constellation and the sun is in the Bull. *For slugs and snails*, collect at a time when they are often most active—when the moon is in the Crab. Ferment for four weeks and apply when the moon is again in the Crab. Burning a quantity of slugs at this time, followed by homoeopathic potentization to D8, is another possible method.

For warm-blooded animals, rodents for example, the skin is the important part, although the whole creature could be burnt. *For birds*, the feathers are taken. The burning is done using a wood fire. Warm-blooded creatures are burnt when Venus is in Scorpion. At this particular time reproductive energies are concentrated, so burning brings about a strongly negative force to which wild creatures are sensitive. These effects may be reinforced when the moon is in the Bull. The ashes are crushed and then 'diluted' using sand or soil. This is then sprinkled around the areas to be protected. Ashes can be potentized to D8, if spraying is preferred to dry ash application.

Readers may be disturbed by these approaches, which are reviewed here to provide a complete picture of measures originating from biodynamics. The first and most obvious point is that no one should carry out practices if they have serious reservations—indeed, if they do, the method may be rendered less effective! Certain issues concerning animals were discussed in Chapter 3 and will not be repeated. Here, we should merely reflect that these methods have features in common with traditional practices and it is above all necessary to bring the right attitude of mind to bear. For those who may regard the use of cosmic timings as weird, it should be said that just because we live and work in an intellectual age does not eliminate their effects. On the other hand, to use them effectively requires reference to a biodynamic calendar which will be discussed in the next chapter.

The use of Equisetum *tea*

This material, otherwise known as horsetail or preparation 508, is used as a prophylactic measure, strengthening a variety of field crops and fruit trees against fungal disease such as mildew and rust. While horn silica acts in a largely non-material way, enhancing forces connected with the silica process in plants, *Equisetum* tea provides a quantitative input of silica. The ash of *Equisetum* is rich in silica. Though not commonly regarded as a plant nutrient, silica is nevertheless recognized as a protector against fungal attack. It inhibits the formation of phenolic toxins in leaf cells resulting from the action of fungal enzymes.

Collect the branching shoots of horsetail (like little pine trees) shortly before sporing. At this time the plant has its highest silica content. Use fresh or dried. If necessary, store the herb in a dry, dark place. Boil in water to make a decoction or tea. Take one part of herb to ten parts of water (two good handfuls in a pan of water), bring to the boil and simmer for about 20 minutes. Allow to cool, strain and dilute one part of tea to two parts of water. Alternatively, if using dried horsetail, 30–50 g is milled and then allowed to simmer for 1 hour in about 1 litre of water. The resulting decoction is then diluted ten times. In either case a brief vortical stirring could be carried out. *Equisetum* tea is mainly used in anticipation of cloudy, damp weather with low light levels. During such periods it should be applied as a fine mist in the morning at weekly intervals.

Animal health

Finally, a few brief points should be made about animal health and disease. We live in difficult times regarding contagious diseases, from avian and swine influenza to foot and mouth disease and blue tongue. If global warming proceeds as a trend, our animals will no doubt continue to experience diseases to which they have not previously been exposed. Whatever ideals are espoused on organic and biodynamic farms, the latter exist within a wider national and global agricultural community. In such a context, to vaccinate would seem preferable to hideous mass culls. This is not to subscribe unreservedly to vaccination for, as with human beings, other problems could result from it.

Nevertheless, susceptibility to disease is a function of the health of the organism and the strength of its immune system. Feed quality and levels of stress are crucial. With biodynamic methods, and in particular the use of horn silica, biodynamic farmers have the capacity to raise the value of feed to the highest level. It is also the case that, following Steiner's final agriculture lecture, we need to become more sensitive to the *dietary* needs of our now highly domesticated animals. In this way we can creatively raise the vitality of animals for the purposes they are intended.[45]

Animals on organic and biodynamic farms are, as far as possible, given homoeopathic remedies and natural products (such as cider vinegar and garlic for internal parasites), rather than resorting to allopathic intervention. The latter inevitably disturb the immune system. To supplement existing strategies there is now a widening range of potentized preparations, mentioned earlier.

6. Working Practically with Astronomical Rhythms

To introduce people to the different aspects of biodynamics I sometimes use the following analogy. To listen to the radio or watch television one requires a receptive instrument. Maintaining a healthy soil and application of preparations represents such basic provision. Tuning to particular stations may be likened to the use of a biodynamic calendar. While this analogy is not intended to have wide appeal it does underline the fact that those wishing to use the calendar must be working organically, and preferably be using the biodynamic preparations in order to achieve worthwhile results. It is no use being interested in astronomy while practising methods that are largely inimical to these finer effects on life processes. I say this, knowing that the use of lunar-zodiacal rhythms is an aspect of biodynamic practice widely adopted by organic gardeners who have yet to embrace biodynamics in the full sense.

Calendars and their purpose

Biodynamic calendars are now produced in many countries, including Germany, the UK, Italy, the USA, Australia, New Zealand, India and Sri Lanka.[1] They are based on standard astronomical data for sun, moon and planets (such as *Raphael's Ephemeris*), placing special emphasis on lunar rhythms (see below). Basic information consists of the positions of sun and moon against the constellations of the zodiac for each day of the month. The moon's other cycles, which include phases, nodes and perigee, are also indicated, as are the more important planetary relationships. Using all this information, recommendations are then made for working with different categories of crop plants, and times which should be avoided are also given. Information in a biodynamic calendar is valid for anywhere in the world, the only essential adjustment being for time zone.

Calendars are associated particularly with sowing and planting but in practice apply to all conceivable horticultural operations relating to each type of crop. Thus we can include cutting and pruning, grafting, tillage, weeding, and the application of liquid fertilizers and biodynamic preparations (but see later comment). Calendar timings have also been applied to compost-making and to the various operations involved in bee-keeping.

You could be forgiven for thinking there was enough work involved with farming and gardening without the further complication of cosmic timings, particularly as the weather or life's other circumstances interfere with many well-laid plans. However, the reason for using a biodynamic calendar is that one can further strengthen the plant and improve its ability to produce the desired crop. Using specific timings focuses or concentrates forces from which the different parts of plants can benefit. This improves pest resistance and yield. It also contributes to produce offering the best quality nutrition, not only in terms of conventional parameters but containing the life forces so essential in our food (see Chapter 10).

A number of annual publications address lunar sowing and planting, some deriving their inspiration from biodynamics, others making no reference at all to biodynamics. One has therefore to be prepared for different approaches and divergent advice in the various guides. There is always a risk in being too heavily dependent on one school of thought, yet since the present book is concerned with biodynamics it will naturally emphasize biodynamic thinking. Even so, as will become evident, it is timely that 'accepted wisdom' is subjected to the eye of scrutiny.[2]

A basic astronomical framework

When making use of a biodynamic calendar it is helpful to have a basic understanding of the movements of the earth, sun and moon, beyond which lie the constellations of the zodiac.[3] In the first place, the earth rotates on its axis to give us day and night. Meanwhile, the earth's annual orbit around the sun defines an elliptical path or plane known as the ecliptic, for objects crossing this imaginary plane can cause eclipses. The ecliptic plane is visualized by observing the sun's passage across the sky. The inclination of the earth's axis to the ecliptic plane (23.5°) accounts for the seasons and for the designation of *tropics* north and south of the equator.

Because the moon and all but one of the planets have orbits which are inclined only slightly to the ecliptic plane, we can expect them to be visible broadly within this same band of sky. Mercury is so close to the sun that we never see it due to the sun's brightness. Venus appears either as a morning star before sunrise or as an evening star after sunset—at other times it is not visible. The more distant planets traverse our sky at any time of the day or night. The earth's progress on its solar orbit results in the sun lying in front of each of the twelve constellations for about one month.

The moon orbits the earth each month. We experience this not by its movement across the sky from east to west, but by its daily displacement towards the *east*. This gradual movement means that the moon lies in front of different star groups or constellations every 2–3 days, completing the entire circuit in 27.3 days (compare the sun taking one year). This is known as the moon's *sidereal* or star cycle. It is remarkable that this is the same period which is taken for the sun to rotate on its axis to exactly the same position as viewed from the earth.[4] Thus, in a sense, solar and lunar months are united. Equally remarkable is that as the earth rotates the moon does so, but at such a rate that it always presents the same face towards the earth. Such correspondences are symbolic and could well be anticipated from what Rudolf Steiner has said regarding earth and moon originally forming one larger body.[5]

While the moon is completing its sidereal cycle, new moon and full moon occur at approximately fortnightly intervals. The latter relates to the relative position of moon, earth and sun. Light reflected from the moon's surface originates from the sun, the sequence of lunar phases relating to the moon's position in its orbit (Fig. 6.1). Needless to say, the brightly lit (convex) side of the moon always faces the sun and therefore points along the path of the ecliptic. New moon represents the astronomical arrangement known as *conjunction*—where sun and moon are in alignment on one side of the earth. At this time, the moon seems to disappear because there is no part of its surface able to reflect to us, while the sun's glare obscures visibility.

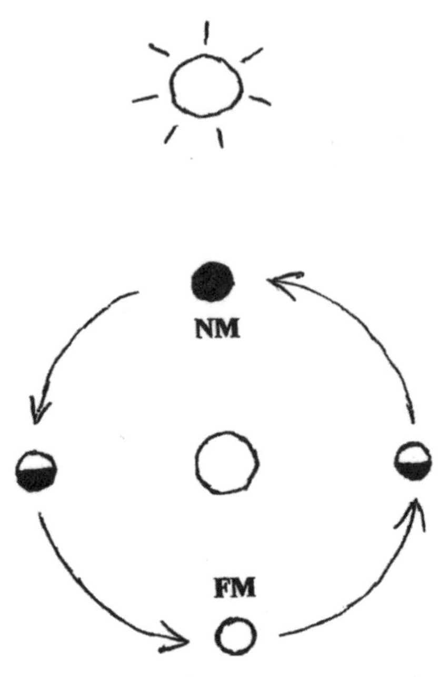

Fig. 6.1 The lunar phase cycle. NM New Moon. FM Full Moon

By contrast, full moon represents *opposition*—where sun and moon are on opposite sides of the earth. At this time, weather permitting, the moon will be visible all night. In addition, it will be observed that full moons are at their lowest elevation in midsummer and at their highest elevation in midwinter. This is because the position of the full moon will always be *opposite* the seasonal position of the sun.

The time from one full moon to the next is known as the moon's *synodic* or phase cycle which is 29.5 days. This is longer than the sidereal cycle because of the earth's movement around the sun in the course of the lunar month. This results in the sun appearing to stand in front of a different constellation and to have 'moved' through one twelfth of the zodiac.

Meanwhile, the moon orbits the earth at 5° inclination to the ecliptic plane, so for part of the month it is above that plane and for part of the month below it. When it crosses the ecliptic plane as it ascends or descends it is at a *node*. This particular movement is referred to as the *nodal* or draconic cycle and takes 27.2 days to be completed (Fig. 6.2). If we consider the phase cycle, we will realize that only when the moon is at a node will eclipses occur. At such a time the moon is not just in alignment, but crossing the earth's ecliptic plane as well. An eclipse of the moon takes place at the time of the full moon—the earth's shadow being cast on the moon. An eclipse of the sun occurs at new moon, when the moon obscures the sun. Variations in the character of eclipses depend on the *degree of coincidence* of these events as well as the *distance* between moon and earth. As the moon orbits the earth it does so elliptically, meaning there is a point of closest approach, perigee, and a point of farthest extent, apogee. When full eclipses occur at apogee they are described as *annular* while at perigee they are *total.* This further cycle of the moon—the *perigee-apogee* or anomalistic cycle—takes 27.5 days (Fig. 6.3). When full moon occurs at perigee one really can detect its increased size!

As already mentioned, the moon moves against the background of stars. As a result of its various motions, but principally the nodal cycle, it displays ascending arcs in the sky for two weeks followed by descending arcs. This is referred to as the *ascending* and *descending* moon. Very slowly in the course of time, the background of stars against which this ascent and descent takes place will change. The moon is currently at its highest elevation passing from Taurus (Bull) into Gemini (Twins) and at its lowest elevation while passing from Scorpio (Scorpion) to Sagittarius (Archer).

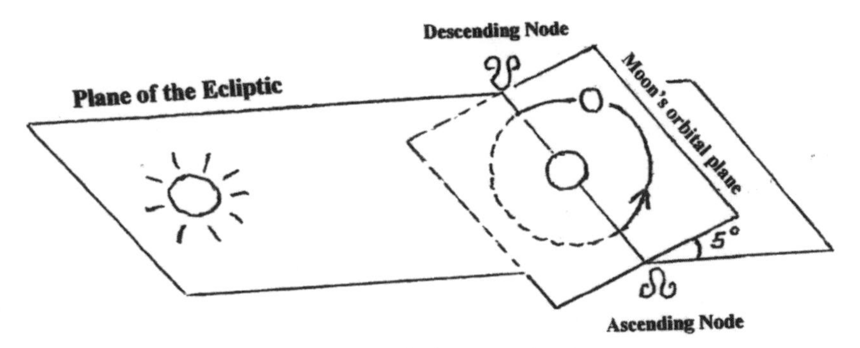

Fig. 6.2 The lunar nodal cycle

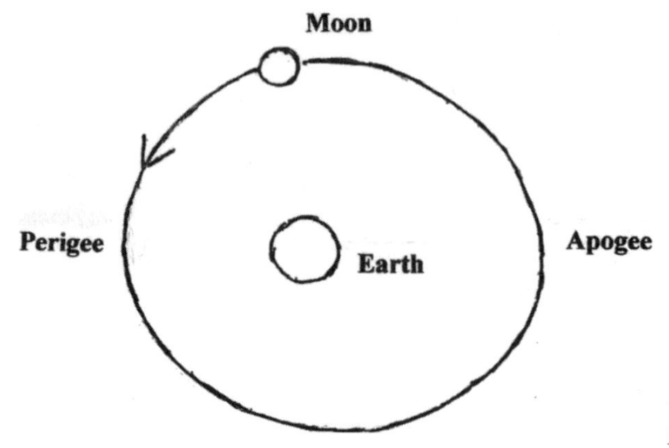

Fig. 6.3 The lunar perigee-apogee cycle

Similarly, the sun's maximum elevations in the southern sky at summer solstice and winter solstice occur against these same constellations.

The moon's phases

The moon's influence upon the earth is expressed most obviously in the tides. We have the highest high tides at full moon and lesser high tides at new moon, while neap tides—where there is minimal tidal range—occur at the points between, where we see a half-moon. There is even some statistical support for increased meteoric dust entrainment in the atmosphere at full moon and new moon leading to higher rainfall amounts 3–4 days later.[6]

The moon has long been recognized for its influence on plant growth through mediating water processes. In Chapter 1, reference was made to traditions of sowing and planting in conformity with lunar phase, the more widespread because, weather permitting, this cycle of the moon is clearly visible. Steiner himself advised farmers to sow seeds during a waxing moon, except in the case of legumes. In experiments carried out in the 1930s, Lili Kolisko showed that vegetables sown before the full moon generally developed into higher-yielding crops as compared with those which began life before new moon.[7] However, the price of such higher yields may well be reduced flavour and keeping quality.[8]

Full moon also has physiological and behavioural impact on animals and human beings. Cows and, especially, goats come into heat at this time while insect activity, most noticeably in the tropics, reaches a peak. Full moon is unsurprisingly a time for more animal births, and childbirths. In fact all body fluids are affected. Trauma cases involving bleeding tend to be more serious and snake bites more lethal around the full moon. In addition, there are difficulties for the mentally disturbed, greater frequency of prison riots and, for many people, greater difficulty sleeping. When all is considered, it is not surprising that community life was at one time attuned to this rhythm.

As the above examples show, lunar influence clearly extends beyond water processes. It connects with reproductive activity and the mental (astral) realm, thus penetrating deeper into nature than might have been thought. For ordinary growth, it seems the earth's own forces are sufficient but for reproductive activity these forces need to be enhanced by the moon's influence.[9] It is clear that the more reflected sunlight we receive from the moon, the stronger this lunar influence becomes—further evidence that light carries more than just illumination. The lunar effect thus reaches a peak at full moon and, by analogy with solar radiation, is stronger when the moon is at higher elevation in the sky.

The constellations and signs of the zodiac

In conventional terms, there are twelve major star groups or constellations arranged around the ecliptic. From ancient times, these have been given animal names, hence the name *zodiac* or animal circle. Although today we simply see groups of stars which can be linked up into patterns, people formerly experienced the spiritual beings or forces which lay behind these and other regions of the heavens. It was pointed out in Chapter 2 that at one time there was an appreciation that from the zodiac came different principles (ethers) which collectively underscore life on earth. Thus were the twelve constellations divided into four groups representing the primal elements, with three constellations—arranged in triangular fashion—representing each element.

Given the evident symmetry of this arrangement of cosmic influences, one has the distinct feeling that if it were to be used in a practical way we should be dealing with *equal* divisions of the ecliptic. As is well known however, in the system adopted by Maria Thun and other biodynamic calendars connections are made with the *unequal* astronomical constellations (Fig. 6.4). These are the actual constellations we see when we observe the sky. Thus if we proceed round the zodiac anticlockwise the elements earth, air, water and warmth are represented in turn.

This is not the place for a detailed exposition of the different zodiacs, nor the controversies which have arisen over allocation of names to parts of the sky.[10] However, we need to be aware that the *astrological signs*, based upon the tropical zodiac, share the same names as the *astronomical con-*

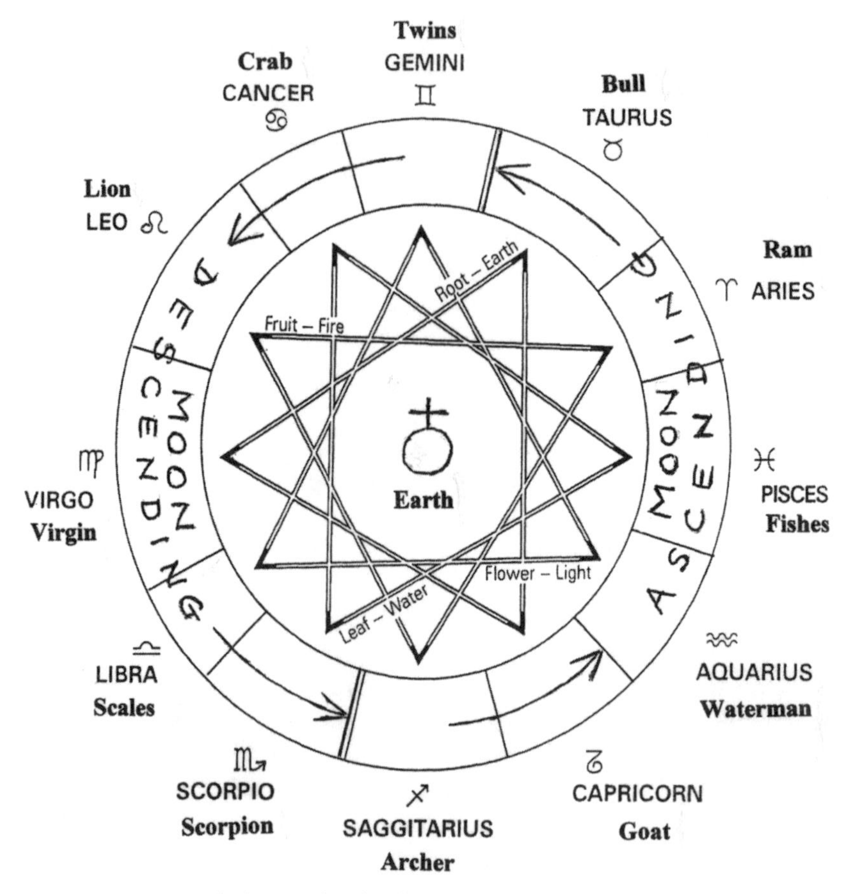

Fig. 6.4 The lunar sidereal cycle as used in most biodynamic calendars (based on a diagram by Maria Thun)

stellations[11] except that the former are normally in Latin while the latter have been translated into a vernacular (Fig. 6.5). The 'signs' are an arrangement, embedded in history, in which each occupies one twelfth of the zodiac—a 30° segment. Originally, in the time of Ptolemy, these two zodiacs shared the same point of origination at 0° east, the 'starting point' of the constellation of Aries (Ram). However, due to the effect of precession, at the spring equinox the sun now rises in Pisces (Fishes).[12]

Plant life is subject to the direct effects of cosmic rhythms, though at this point we should avoid being dogmatic about whether this relates to the unequal constellations or with very similar, but equal, divisions of the circle. In animals and humans, these rhythms are internalized in the physical organism through mediation by the planets.[13] Yet the con-

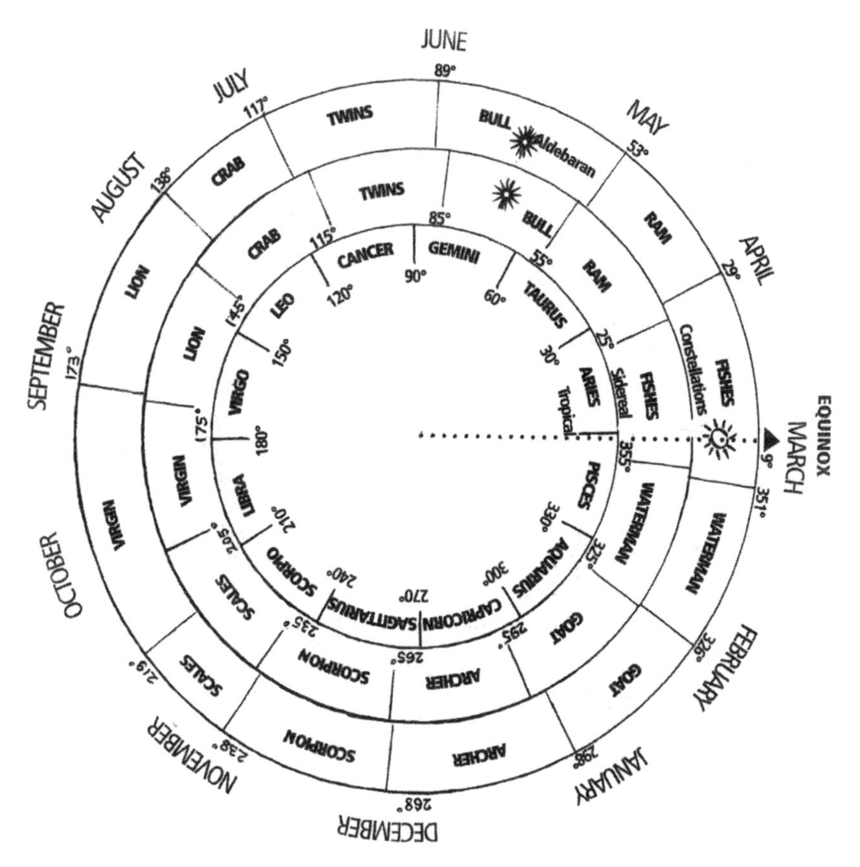

Fig. 6.5 Correspondence of different zodiacs (from Ian Bailey, 2008)

stellations still represent real influences on the sentient or astral body.[14] It was for this reason that animal representations arose, and it is this connection which particularly relates to astrology. However, when we consider the higher spiritual part or individuality of the human being, this is said to connect with regions even further out in the cosmos, so human beings are ultimately not limited by the zodiac.[15]

The sidereal moon

From the beginning, biodynamics has been concerned with establishing balance among the various forces governing plant growth. The preparations are measures to achieve this. Steiner warned against practices which would over-stimulate the lunar force. Although different workers have researched the effects of the moon,[16] biodynamics has never actively promoted the use of the waxing and waning moon in agriculture. Despite this, there is a persistent tendency in the literature, especially with more popular publications and the press, to refer only to the moon's *phases*.

Although he never laid out a scheme for a calendrical system, Steiner did drop certain hints. In many lectures he spoke about the influence of moon and planets on plant growth. Notably he stated that 'along with the moon's light the entire reflected cosmos comes toward the earth'.[17] We are also made aware that the sun's quality alters as it appears in front of different zodiac constellations and that it can be strengthened or weakened by planetary relationships.[18] The fact that the moon passes these same constellations within the space of one month provides, therefore, a very reasonable template for research. Such investigative work has been conducted for over 50 years by Maria Thun in Germany, assisted in recent years by her son Matthias. She was by no means a pioneer in the field of calendar publication but her early results were sufficiently encouraging that she began in the 1960s to make the recommendations which are now taken for granted in her annual biodynamic calendar.

Maria Thun's early research was based on sowings of crops with the moon 'in front of' successive constellations in the course of one month, with trials repeated in subsequent years. Here, we will not attempt a critique of Maria Thun's methodology but merely report its key findings.

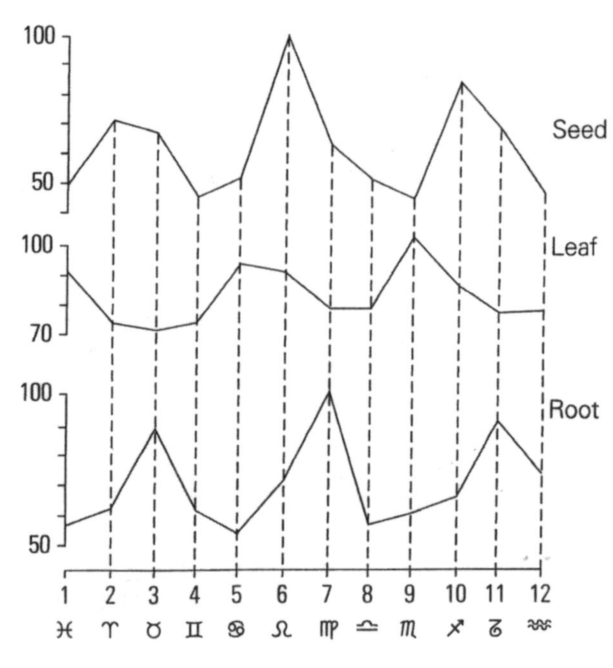

Fig. 6.6 Relative yields (% of maximum) for different components of radishes planted with the moon in successive constellations (after Thun and Heinze, 1973)

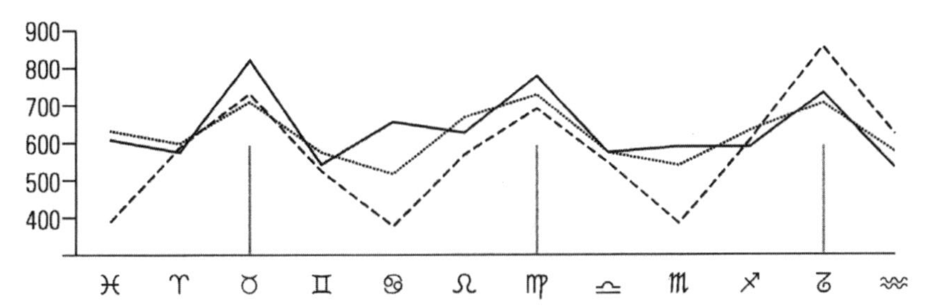

Fig. 6.7 Yields of potatoes (average grams per plant) when planted with the moon in successive constellations. Three separate years' experiments __ 1965 _ _ 1968 ..... 1982. (After Thun, 1983)

To this day, one of her most revealing trials shows the results of sowing radishes in successive lunar-zodiacal periods. After equal lengths of growing period, the radishes sown at earth-constellation times had achieved highest yields of radish, while at other times relatively more leaf, flower or seed was achieved, according to the constellation (Fig. 6.6).[19] Maria Thun thus demonstrated that plants can be manipulated morphologically by choice of sowing or planting date—in other words, the sidereal moon has a tendency to affect the partitioning of materials within each crop. Also illustrated is Maria Thun's work with potatoes (Fig. 6.7) and that of Nick Kollerstrom on broad beans (Fig. 6.8).

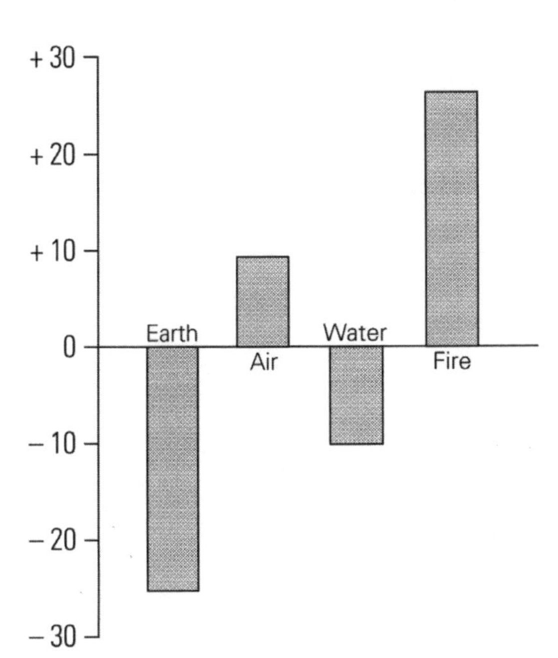

Fig. 6.8 Yields of broad beans according to lunar-zodiacal sowing. Data are percentage deviations from the mean for each group of constellations. Results for 1976 and 1977 have been aggregated. (After Kollerstrom, 1985)

In temperate regions there is a recognized seasonal intensity of growth which manifests as a growth curve over the course of the year, but with sowing dates confined within a single month this factor is not normally an issue. However in certain cases where researchers have failed to achieve similar results to Maria Thun, it has been shown that applying a seasonal growth function enables some reconciliation.[20]

So the calendar system developed by Thun is based on the moon's *sidereal* cycle—its passage in front of the visible constellations. In Fig. 6.4 the earth is portrayed in the middle while the moon circulates anticlockwise. Each constellation may be viewed as having its inherent forces channelled by the moon—forces which pertain to the different parts of plants. Meanwhile, for half the month the moon ascends in the sky, and for half it descends (see below).

Crop categories and timings

It is obvious, but ought nevertheless to be stated, that to make use of the sidereal moon requires a clear understanding of the nature of each crop plant. What are the organs required for harvest and therefore how is the plant to be classified in relation to the four elements already mentioned? The calendar designates times appropriate for root crops (earth element), flower crops (air/light element), leaf crops (water element) and fruit (or seed) crops (warmth element). In most cases a child would be in no doubt as to which category each crop belonged but sometimes it is not quite so straightforward.

Let us take root crops to begin with. The potato is strictly a tuber or underground stem, yet the fact that it is a crop maturing underground makes it fall into the root category. Other temperate roots would, of course, include radish, carrot, swede, fodder beet, parsnip, celeriac and, less obviously perhaps, onions. While the latter ripen on the surface, they are effectively regenerative or storage organs and as such align themselves with other root crops.

Leaf crops include the whole range of salad leaves, together with spinach, chard, celery, kohlrabi, kale, cabbage and Brussels sprouts. And despite their relation to the onion family leeks are grown mainly for their stem and leaf characteristics. Flower crops are not limited to those grown as ornamentals or as companions for insect control. They include broccoli, calabrese, cauliflower and the globe artichoke. Choice of the flower day is also appropriate for oil-bearing plants such as linseed, rape and sunflower. Herbs grown for their essential oils are functionally leaf crops, but since quality rather than bulk of production is a priority, the light or warmth element might well be explored.

Fruit crops are extremely diverse. They include apples, pears and cherries (top fruit) along with currants, gooseberries and strawberries (soft fruit) mostly grown for dessert purposes. Other vegetable crops which we gather for their 'fruits' include different varieties of beans and peas (Leguminosae), varieties of aubergines and tomatoes (Solanaceae), together with marrows, courgettes, cucumbers and pumpkins (Cucurbitaceae). The latter two categories are frequently grown under cover. Cereals (Poaceae) are clearly also fruit crops despite our use of the word 'grain' and its association with 'seed'. According to these principles, fruit days (warmth element) will therefore be best for quality and yield of fruit and seed crops. This should not be confused with the wider issue of seed development which is discussed in Chapter 7.

It is likely that beneficial effects based upon timings will be reinforced if other necessary crop management operations, such as hoeing or weeding, are carried out in a similar constellation. Such are the complexities of real life that this may only occasionally be possible. However, it is important to realize that by this means it may be possible to compensate for having initially missed the appropriate sowing or planting time for a particular crop.

The calendar can also be used for general estate management—gardens for most of us! For optimal productivity, forest or timber trees will have a relationship to the vegetative process and therefore to water and leaf. For lawns we have different options. Grass is clearly a leaf crop but how frequently does one wish to be cutting it? Cutting grass on leaf days will cause it to grow more strongly. Cutting on a root day will help it survive drought, while cutting on flower days, as the writer has discovered, encourages dicotyledons. So, providing a judicious mowing height is maintained, a wild flower lawn can become a reality. Similar principles can be employed with hedge cutting. To encourage a young hedge, leaf days, while to discourage too much growth, perhaps root days would be better.

Use of the ascending and descending moon

The German (Thun) and New Zealand calendars recommend use of the descending moon for transplanting, this being highlighted in the calendar by the terms 'northern and southern planting times'. Others, including the American *Stella Natura*, make no stipulations in relation to the ascending or descending moon. The main problem with using restricted *planting* times rests on practicality. The temperate growing season is limited and, with meteorological and personal issues to be navigated, cutting the available planting time by half is not particularly helpful. In the UK, I have even discovered people avoiding the ascending moon period *for sowing*, simply because it is accompanied by the term '*southern* planting time'! Nevertheless, there can be a sound basis for using the descending and ascending moon if circumstances permit.

Having noted the appropriate zodiacal days for particular plants, we should consider the impact of the lunar process. Let us take two contrasting situations. When transplanting, the re-establishment of a root system is the priority so this is best done in the *descending period* when the moon's effect on plant water uptake is diminishing. Pruning too, having selected the appropriate constellation, should be done on the descending moon which will reduce the risk of disease attack. Use of the descending moon could also connect with hedge cutting or lawn mowing. On the other hand, with sowing seed, water uptake is essential so the *ascending period* is optimal. Again, with grafting, we need to select a time when the young scion is most likely to be provided with flow of sap. This will be when the moon's effect is strengthening—during the ascending period.

How have the preparations been considered in relation to these lunar movements? Following Maria Thun, horn silica might be applied during the lunar ascending period when there is a stronger upwards movement of water. This mirrors the normal early morning 'out-breathing' time for its application. For horn manure one would choose a descending moon. However, we must be very clear that the horn preparations relate to a *solar diurnal rhythm*. To look for further auspicious times is, in this writer's view, a recipe for not getting the job done in a timely way. The priority should be to choose the *zodiac time appropriate for the crop being grown* rather than 'lose the plot' by focusing on an ascending or descending moon—after all, full moon also occurs, and this may happen in the descending period! Equally, some might argue for spraying horn silica on a light/air day (horn manure on an earth day) but here again there could easily be conflict with the impulse which should be given for individual crops. Each grower will want to use the calendar in their own way but let it not end up being a source of uncertainty and confusion.

Additional uses of the calendar

Generally speaking, animal husbandry lies outside the scope of the calendar.[21] This is not surprising, for the animal world is not so directly subject to the zodiac as are plants. Furthermore, the care of animals involves constant daily routines so is not susceptible to such a system. However, Maria Thun reports that bees respond to the different lunar-constellation elements. In their different activities—gathering nectar, gathering pollen, honey processing and comb building—there appears to be a relationship respectively with the elements of warmth, light, water and earth. That this gives the beekeeper clues to the current priority in the hive is valuable. It enables us to be more sensitive to the activities of bees and their accompanying elemental spirits at this critical time. It is also evident that micro-organisms respond to cosmic rhythms, so bakeries may well experience variations on a day-to-day basis (see below). It is also reported that milk processing for butter and cheese is affected by the lunar-zodiacal constellation.[22]

One can always take a good idea a little too far, and some will wonder about using this system for compost making. We shall juxtapose different views. On the one hand compost consists of plant and animal remains reduced to an earthy condition and as such may seem to best connect with that element. But it is possible to view the compost process as one of ripening, in which case warmth days would seem the most appropriate. That compost requires *all* the elements working in combination is also pretty obvious!

How does the lunar-zodiacal process actually work?

In view of the fact that cosmic forces appear to be conducted by the moon, one is entitled to wonder how this process works. Reference has been made to the coincidence of the lunar sidereal cycle with that of the sun's axial rotation. Maria Thun wrote that the moon 'is a kind of pointer to the rhythm of the sun facing different parts of the zodiac ... [it] becomes a transmitter of cosmic impulses to which the plant reacts, as do soil organisms ...'[23] Indeed, despite its strong connection with water, the moon self-evidently transmits *all four* of the elementary conditions or ethers.

It is recognized that cosmic alignments have a strengthening influence and it is probably right to think that a kind of channelling is occurring. We see the moon as a solid, lifeless object but in spiritual terms—in relation to the ethers—it may be virtually transparent. We might then visualize the moon acting like a condenser lens, concentrating particular zodiacal forces which would otherwise be scattered. We should also note that whenever the moon is in front of a 30° zodiacal sign it is automatically in a 120° (triangular) relationship with two other similar signs, this being likely to strengthen the element relationship.

A further and vital question may be asked as to why it is only the zodiac constellations we refer to, while the rest are apparently ignored. In one of many remarks concerning the broad span of the cosmos, Steiner commented that '*everything* there is by way of stars in the sky has a definite influence on the earth as a whole and specifically also on the human being.'[24] Clearly then, the vast majority should *not* be ignored. If we consider the issue of alignment, zodiac constellations can be seen to have a special significance. This is because they are traversed by sun and moon which therefore draw in forces from the zodiac behind them. Of the zodiac constellations, it was said that 'the space that lies before us as our visible universe is divided in this way. *The signs merely denote the boundary of a certain section of space.*'[25] How then, might we pursue this question?

Let us be aware that the visible cosmos was always thought of as a *celestial sphere*, the equator of which is a projection of the earth's equator. The planets too, as we mentioned in Chapter 2, were at one time felt to have influence within their 'spheres'. In the same way that the earth has meridians which are united by their temporal relationship to the sun, it seems likely that constellations above and below the celestial equator are similarly united. As such, they would best be able to channel their forces towards the earth when sun or moon is in 'meridional alignment' with them. When we consider the vast array of classical names given to the entire starry heavens, such an interpretation seems wholly justified. If this hypothesis were to be substantially true, the zodiac constellations would have the noble task of coordinating the spiritual impulses from one-twelfth of a sphere, rather than presiding over unequal, and to some extent disputed, segments of a circle.

Choice of zodiac is a key issue

At this point we should note that Kollerstrom, in his research on zodiacs, finds no historical evidence of the use of the zodiac in the way that biodynamic researchers since the 1930s have preferred to interpret it. Maria Thun has claimed that her effects relate to the visible and unequal constellations. However, Kollerstrom's analysis of her data indicates that while there is certainly a 'Thun effect' against the constellations, it is in fact the moon's movement against 30° 'signs' which appear to show best fit. These, however, are not the 30° signs of the tropical zodiac as originally used in astrology, but the latter *adjusted to account for precession.*[26].

It is this transformed 30-degree system relating to the star patterns we actually see, and which appears to have been used in antiquity, which Kollerstrom refers to as the *sidereal zodiac* (Fig. 6.5). Farmers and gardeners therefore need to be aware that the constellation framework offered in their biodynamic calendars should not be regarded as 'written in stone'. While greatly respecting the role that Maria Thun has played in bringing the four-element system into our modern consciousness, the advice offered here is to use the 30° *sidereal zodiac* as a more reliable template for achieving plant response. Indeed, as the previous argument has suggested, this view allows for a three-dimensional and more spiritually inclusive view of the universe. As will be seen, there is very substantial overlap of the two zodiacs—the difference being the beginning and ending times for solar and lunar transit of each sign.

Understandably, many adopt the Thun system as an act of faith, and the fact that it has persisted for so long in its present form is due in no small measure to the limited numbers who conduct investigations, however simple. To encourage practitioners to be independent-minded, Ian Bailey points out there is a great need for observational and perceptual work in this field.[27]

Planetary relationships or aspects

The question of planetary influence is more complex than can be covered in this chapter but suffice to say that each planet exerts direct and indirect influences on life processes, akin, for example, to breathing in and breathing out or to nutrition and excretion. And because of constantly changing cosmic relationships the normal influences of each planet may become obscured or accentuated.[28] Steiner referred to the sun's light being strengthened or weakened by planetary influences.[29] When planets are in *opposition* to the sun, moon or another planet they augment or ameliorate particular effects. Most often they lead to an intensification of the life processes of the plant.

The planets have been considered to embody one or other of the Aristotelian elements. Thus are Saturn and Mercury connected with warmth, Jupiter and Venus with air or light, the moon and Mars with water, the sun and earth with the earth element. In this way the planets are connected with the cosmology of the solar system. However, so far as can be judged here, such a conception appears to have arisen more from Steiner's influence than by tradition. Nevertheless this offers insight into how individual planets mediate cosmic forces originating from the zodiac. Looking again at the astronomical zodiac (Fig. 6.4) it will be noted that oppositions link either light and warmth constellations or earth and water constellations, thus benefiting different crop categories. There are therefore different opportunities for plant husbandry arising from careful use of the calendar.

On the other hand, planets in alignment with each other are likely to pose problems. Astronomers call this *occultation*. The fact that one stands in front of another as viewed from the earth means that the influence of the one obscured will be negated. Under these circumstances, despite the fact that the phenomenon quickly passes, Maria Thun reports that non-germination or very poor growth results. Seed viability is not always consistent despite apparently good pedigree. This may be explained by such a conflict arising at a crucial stage of the parent crop.

With perennial plants in particular, such as fruiting or forest trees, planetary relationships are helpful. The most dependable of these is Saturn in opposition or triangular (120°) relationship with the moon. The Greek poet Virgil referred to Saturn as the 'gardener's friend'. Although there seems little doubt that the rhythms of particular planets play strongly into the life of different plant families, it is important to realize that *all* the planets affect different stages of the growth and fruiting process.[30] For instance, Maria Thun has been able to develop six different coloured Mirabelle plums from the same stock simply by allowing the ripened plum-stones to germinate under different planetary aspects.[31] Renowned herbalists such a Nicholas Culpeper (b. 1616) have always recognized planetary 'rulership', yet over-adherence to this view would appear somewhat dogmatic.

Planets may also influence the weather. Maria Thun claims to have shown that when a planet associated with a particular element stands in front of a similar constellation the constellation's effect is intensified, while if the planet is in front of a constellation with a different effect the planet's effect is neutralized.[32] For this purpose the outermost three planets are brought into the picture. This is no simple matter to contemplate, for any forecast should have consequences for the planet as a whole. Moreover, primary effects are likely to be obscured by terrestrial factors, inducing a lag effect, so that detection will always be problematic.

Times to avoid

It is axiomatic that one should adhere to the appropriate times for the purpose intended—earth constellation for root crop, water constellation for leaf crop etc.—but it is less obvious that there are times one should *avoid*. This applies to the 2 hours at the start or ending of each zodiacal transit by the moon. These are transitional times and there may be further reason for caution in view of the way that international time zones are allocated. It is also important to avoid particular times such as full moon, moon at perigee, moon or inner planets at their nodes, as well as eclipses. The latter have particularly marked impact on natural processes. They are times when the lunar force is unbalanced or when the flow of energy across the earth's ecliptic plane is disturbed. In such cases, calendars are

Fig. 6.9 Samples of horn manure treated by the circular chromatography method. Left: At solar eclipse. Right: The following day. (F. Schikorr, unpublished)

united in recommending extended periods when agricultural operations should be avoided, for example 12 hours either side of a lunar node or perigee, and similar times for Mercury or Venus nodes.

During the solar eclipse of 29 April 1995, equal portions of biodynamic 500 were extracted in a standard amount of solution which was then allowed to spread radially by capillarity across circular chromatographic paper[33] (Fig. 6.9). Despite extra time, the test conducted during the eclipse failed to spread across the paper, while less than 24 hours later a normal expression of concentric and radial patterns was resumed. Without change in ambient conditions it is concluded that processes of physical movement are impeded at such times, let alone expressions of life energies. In view of the special nature of water (Chapter 8) one should not be surprised at this.[34]

Maria Thun's experiments have shown the adverse consequences of sowing or planting at all such problematic times. For instance, while lunar perigee leads to failure or stunting of growth, apogee, otherwise much less of a hazard, tends towards woodiness in crops like carrots. Here we appear to have the two extremes, one with too much water influence and the other with too little. While the nodes and perigee can in this way lead to a significant weakening of plant growth and susceptibility to pest attack, there is also evidence that they affect baking, so micro-organisms appear to be sensitive to the same disturbances.

Harvesting times

Traditions are revealing on this subject. Pliny stated that wood rots more quickly when harvested at full moon, while native people of the Amazon basin would always cut thatch for the roofs of their huts at the new moon, otherwise destruction would be rapid. Today we talk of *keeping quality*, and the point has been made that crops starting to be grown at full moon may achieve higher yield but keep less well. These observations reinforce the notion that timing confers a predisposition on living materials—rather like a horoscope. So for produce to remain fresh for longest, if using the traditional moon phases, it should be harvested late in the waning moon or at new moon.

When working closely with the biodynamic calendar, the general rule to follow is to use the element days appropriate for the crop, especially if you are to consume the produce straight away. For example, Maria Thun reports that flowers have a stronger scent and remain fresh for longer if cut on a flower day. For leaf crops that are cut on a regular basis it seems unreasonable to specify restricted times of harvesting. On the other hand, there is an argument for avoiding leaf days in order to improve keeping quality, a situation also favoured by new moon or moon at low elevation.

There is also the possibility of taking the daily breathing rhythm into account. Hauschka showed that harvesting first thing in the morning ensures food will contain more vital force[35] (Chapter 10). Meanwhile the grower concerned to maximize weight of produce would look favourably at increased water content, and consequently harvest at full moon!

Further thoughts on the use of the calendar

The purpose of a calendar is to enhance cropping yields and quality. That is fairly obvious. But to make effective use of a biodynamic calendar you need to plan sowing in particular, as far ahead as possible. Clearly some coordination with the family diary is necessary. Records of sowing dates from previous years will be valuable as a reminder, even if they are only a rough guide. A sure recipe for disillusionment is a continual failure to get organized in this way.

And yes, having bought a copy of a calendar in a temperate region, don't be too disappointed that you can only make effective use of it for 6–8 months. On the other hand, daylight hours are much longer than in the tropics where, unless working with tractor headlights, one needs to mark the calendar with the hours when it is light enough to work!

Calendars should never become a cookery book. They only contain recommendations. And if you want to ignore the advice, do so, but at least find out what hazards threaten if you do! As with the biodynamic preparations, regular use of the calendar is a way in which one can consciously relate to natural rhythms and the working of elemental forces. If it helps in becoming more involved with astronomy, obtain a star map or 'planisphere' and take an interest in observing the night sky. Regrettably this won't just depend on weather, for visibility of low inclinations in particular will often be blighted by buildings or light pollution in built-up areas.

For those working organically, some effects can be achieved using a biodynamic calendar alone, but for best results the lunar-zodiacal timings should be used in conjunction with the biodynamic preparations, as these increase the sensitivity of soil and plant. While a soil with good organic matter content provides the correct foundation, Maria Thun has always stressed that the soil should not be over-fertilized (with raw manure or compost) or over-watered. Well-rotted compost should be used whenever possible. Failure to observe this principle will simply encourage too much purely lunar influence, increasing water uptake and obscuring finer influences from the zodiac.

In this connection it is worth a final comment on protected horticulture—glass or polytunnels. One should not be dogmatic, but due to copious amounts of imported compost and regular irrigation *the prognosis is that plants will be less able to respond in a sensitive way to element timings.* Some might also question whether the type of covering and the quality of the resulting light is a further impediment. Aside from good technique and vigilance, this would appear to underline the importance of repeated applications of the biodynamic spray preparations and ensuring that all compost has been well treated with the compost preparations. All these measures are designed to guide the working of cosmic forces.

7. Seeds: Nurturing a Vital Resource

Overview of the current position

Seeds are horticulture's most vital resource. In times past, farmers saved and exchanged their seeds so that crops were honed to particular districts and soils, as were breeds of cattle, sheep, pigs and poultry. This led to the diversity which was a feature of indigenous, pre-industrial farming. We now live in an era where the greater part of food plant genetic resources has become the possession of commerce and where over 70% of all vegetable cultivars are hybrids.[1] Furthermore, the genetic erosion drawn attention to in Chapter 1 is reinforced as seed production is concentrated on fewer companies.[2] The issue of the seeds market is a vexed one, for every packet of commercial hybrid seed we buy contributes to investment in unwholesome breeding methods and to further depletion of open-pollinated varieties.

Regulatory bodies require that seeds used in organic production systems are derived from certified sources, so a key requirement for the world's organic movement is the availability of seeds produced by organic or biodynamic husbandry. In Britain we are fortunate in being able to source organic seed from our own suppliers and those of the near continent,[3] but demand often outstrips production. This might involve a particular carrot variety or a grass seeds mixture for starting a ley. In particular circumstances therefore, conventional seed, or a proportion of conventional seed, may be permitted by organic certification bodies (by derogation), providing it has not been treated with fungicides or other chemicals.[4]

There have sometimes been complaints that available organic varieties are either not right for the grower's local conditions or the market they are growing for. While in the short term this might restrict growers to varieties that are less suitable for their needs, the strict organic standard should act as an incentive for seed breeders and lead to varieties being chosen that perform better under organic management—even to land races that are ideally suited to local conditions.

Over recent years a great deal of effort has been made in this direction, with programmes designed to support plant breeding and seed production, as well as to protect rare varieties.[5] Notable among these has been the Biodynamic Association's initiative in the UK, centred at Stormy Hall Seeds, Botton Village, Yorkshire, and involving many out-growers.[6]

Later in the chapter we shall outline procedures for the development of seeds for organic and biodynamic production. In the meantime, it is important to build an understanding of the nature of seeds and of how they are formed through various plant breeding methods. We may then be better able to form judgements about the sort of plants preferred for our food crops.

Biological and biodynamic perspectives of seed formation

One can think of seed formation as either the terminus or the culmination of a sequence of life processes in flowering plants (angiosperms). Whether we speak of annuals or perennial plants, this conclusion of the annual cycle is the same. While flowering plants commonly have both female and male parts present in their flowers, a variety of strategies are frequently deployed to prevent self-pollination.

Fertilization is effected when pollen from another plant of the same or compatible species is transferred to the stigma by wind or insect—in fewer cases self-pollination occurs. A pollen tube develops and finally penetrates the ovum at the base of the style. As a result, fertilized cells develop into the seed while the outer carpel formation, depending on species, forms the fruit.[7] More precisely, one of two sperm cells fertilizes the egg cell, forming an ovule or embryo. The other sperm cell fuses with two nuclei in the centre of the embryo sac, eventually forming the nutrient-rich tissue of the endosperm. Taking the example of rice, a cereal grain, egg cells begin development within 12 hours of fertilization while an embryo is distinct after 10 days. This has fully matured and ripened into a seed after 25–30 days. The endosperm has a milky appearance after 8 days, that of soft dough after 14 days and has hardened after 21 days.

Meanwhile, what we normally consider to be the process of fertilization by pollen is referred to in the *Agriculture Course* as part of a warmth process by which an earthly blueprint of the plant is driven into the seed.[8] This blueprint is derived from the living tissues of the plant—the cambium of perennial plants. The physical *form* of the next generation of plants is thus encapsulated by the seed. In the biological picture presented above, the female organs of the parent plant are clearly the recipient (or vessel) for the image of the plant which will be embodied in the seed. The male cells arriving from another plant help us to visualize a force by which this image can be concentrated. Fertilization and seed development (the termination of vegetative growth) are therefore all part of a 'warmth process'. This, of course, is the polar opposite of a 'water process' which we would associate with vegetative development.

The highly complex and concentrated substance of the seed, comprising proteins and other substances, reaches a point in its ripening where it becomes open to influences from the wide expanses of the cosmos. The seed—or actually the embryo—is thus considered to present a minute world of 'chaos' for the working of these outer forces. It is in these outer realms that exist the archetypes of plants, the spiritual beings connected with the different plant families. So in this perfected earthly form—the seed—a vessel has been created for the working of the appropriate cosmic *forces* for sustaining the plant in its next generation. Thus: 'The mineral-like seed remains only an anchorage enabling the being of the plant to find its way back into the world of appearances when conditions warrant'[9] (see seed germination below). According to Steiner, *any new organism, plant or animal, can never arise without such a sequence of processes occurring.*[10] Here, then, we have a picture of how true vitality and viability arises, and we should try to hold this idea in mind when considering more modern methods of plant breeding.

Open-pollination

This term is applied to all varieties of seeds that breed true—plants that pollinate freely and openly with their own kind and that produce a crop resembling the parent. We can say that this process, in which plants do not accept their own pollen for propagation, has led to a continuous exchange of genetic characteristics. Although the origin of many of our crop plants is shrouded in mystery, open-pollinated varieties have in general come about through natural genetic mutation and a combination of cross-pollination and selection. These processes have led to cultivars that expressed desirable traits over the course of our history. The latter include adaptability to climate and soil, disease and insect resistance, taste and storage characteristics.[11]

In most instances plant breeders cross-pollinate plants of a single species to achieve favourable combinations of alleles, but breeding of two species within the same genus is also possible. In addition, members of two different—but not distant—genera have also been crossed. The case of wheat and rye combining to produce triticale is an interesting example.

Researchers began crossing wheat and rye in the late 1800s but fertile triticales capable of producing viable seed were virtually unknown until 1938. In that year Arne Muntzing produced fertile triticale by treating wheat-rye crosses with colchicine. As this induced a doubling of the chromosomes, normal reproductive pairing and division of chromosomes could occur at meiosis. Triticale became a new crop species similar to but distinct from wheat, rye and other cereal grains. Once created and reproduced, a triticale does not revert or break down to its wheat and rye components. Triticale is primarily a self-pollinated crop like wheat, although as a result of its rye component it out-crosses more frequently than wheat which might not be regarded as a favourable trait.

Hybridization

The interbreeding or crossing that has just been referred to is distinct from hybridization, which characterizes all other plant breeding. Hybrids are attractive for their vigour and for the uniformity of their characteristics, which is what major food suppliers demand.

Hybrids have been formed in different ways. For many years they were produced by inbreeding of two original varieties followed by their combination—a process which would never occur in nature. It involves plants having to accept their own pollen, leading to a narrowing of the genetic composition of each variety until desired traits have been selected. The combination of attributes from each parent then gives the commercial breeder scope for marketing a distinctive variety.[12]

Hybrids have also originated from artificial mutation and tissue culture methods. Early in the twentieth century it was discovered that high-energy particles, X-rays and gamma rays, induced plant genetic mutations. Chemicals, too, were employed. From the population of treated plants, selection of potential new varieties could take place. Tissue culture is now widely used in plant breeding. This is a technique for growing cells, tissues and whole plants under sterile conditions in artificial nutrients. The particular strategies include cytoplasmic male sterility (CMS), cytoplasmic male sterility protoplast fusion, ovary-embryo culture, *in vitro* pollination and anther-microspore culture. Hybrid seeds for use in *organic agriculture* may originate from any of these techniques with the exception of CMS protoplast fusion. *For biodynamic growing*, use of protoplasm and cytoplasm fusion techniques is prohibited.[13]

While hybrid vigour—heterosis—is one outcome of hybridization, the narrowing of genetic character is a weakening influence on the cultivar. In consequence, the crop requires more intensive protection from environmental stresses. The result has been that the performance of new varieties rarely meets expectations. This will not be helped by continued deterioration of soils and failure to provide the increasing inputs of fertilizer and pesticide required. Failure to maintain vigour may also be affected by the process of plant breeding being over-dependent on artificial environments and vegetative multiplication. Whatever the explanation, this has played its part in entrenching poverty among many of the world's poor.

A question sometimes asked is whether it is possible to save and subsequently sow seeds that have been harvested from hybrid plants. The answer is that we certainly can, but in practice we don't. Seed deriving from F_1 (first filial) hybrids, where, for example, two inbred lines may have been crossed, does not breed true. F_2 plants germinating from these seeds display a pattern of characteristics with roughly 50% like their F1 parent and the remaining 50% like either of the parents which combined to form the F_1 hybrid. If we were to take this process further to the F_3 generation there would be a complete mixture of types which would be even more unacceptable to the farmer. For this reason, to grow a marketable hybrid crop one is obliged to purchase fresh seed each year.[14]

Transgenic technology

Genetic engineering comprises a range of techniques from molecular biology by which the genetic material of plants, animals or microorganisms is altered in ways or with results that could not be obtained by normal methods of fertilization and reproduction. The techniques include: (i) use of a biological vector, usually a virus, to introduce a foreign gene into plant DNA; (ii) cell and protoplast fusion where live cells with new combinations of heritable material are formed through fusion of two or more cells; and (iii) micro- and macro-injection, encapsulation, gene deletion and doubling, where direct introduction of externally prepared, heritable material is introduced into the cell. Manipulative techniques used in cell biology, including *in vitro* fertilization, transduction, conjugation, natural hybridization and polyploidy induction, are not classed as genetic engineering.[15]

Human beings have been modifying organisms ever since the first steps were taken to domesticate plants and animals. Those involved with this technology are therefore prepared to argue that genetic engineering is merely the latest in a series of developments involving genetic and selective techniques. The creation of HYVs might be cited as a case in point. However, the situation cannot be looked at in quite so simplistic a manner, for there are fundamental differences in approach. We are here involved with taking the creation of new organisms further away from any process that would be likely to occur in nature for which checks and balances have been established in the course of evolution.[16] And with genetic modification comes the possibility to prevent germination of seeds resulting from the first generation crop, otherwise known as 'terminator technology'.[17] This reinforces the control of the seed producer, and in interfering with the natural way in which seed proteins are formed it takes an uncalculated risk with human health.

In contrast to earlier plant breeding where whole organisms or cells were used and where crossing of the boundary of genera was uncommon, the essence of GM is direct intervention with cell contents and the use of genetic material from any source likely to deliver a profitable outcome. In common with most of the current approach to hybridization, this lacks reference to a picture of the whole plant and its wider relationships. It amounts to playing a game of Lego with the living world, for genetic structures are earthly sensors for energies of cosmic origin. It might therefore be suspected that all laboratory-generated organisms will be more strongly connected with the earth and will, in consequence, suffer reduced vitality. On this basis it could be doubted whether the performance of such seed, any more than earlier HYVs, will be maintained through repeated annual production. There is already evidence that predictions of yields and pest resistance by GM crops have not been fulfilled. More rather than less pesticide is being used and increasing numbers of farmers are expressing disillusionment.[18]

A further issue relates to the insertion of animal genes into plants. As was mentioned in Chapter 3, in addition to a purely physical organism animals have both a life energy and a sensory-consciousness. Plants have only the life principle developed fully. For them, what we may call the sensory world (astrality) surrounds them. Spiritual-scientific research has revealed that whenever the astral is drawn into the body of a plant, the latter imparts poisonous attributes. The tendency for seeds to contain toxins can be viewed as a result of the penetration of astral forces. If animal genes, which have a dependence upon or have originally evolved in connection with an astral body, become part of the composition of every cell of a plant, we should be concerned about the consequences. The least we could imagine is that this may interfere with cosmic connections the plant needs to make.

Furthermore, it would be naive to assume that a single such gene in a plant would cause no harm to the animals or humans who depend on it for food. There have been reports of serious allergic reaction to GM foods, while one study appears to show that pigs fed GM potatoes developed lesions in their intestinal tract—the more worrying because of the similarity of their digestive systems to ours.[19]

The Achilles heel of the genetic engineer is the basic truth that you cannot simply add or subtract one gene without affecting the overall resonance of the DNA. This is where the analogy with Lego breaks down and why such experimentation has repeatedly produced undesirable outcomes. Consider, for example, the Flavr Savr tomato designed for longer shelf-life, the flavour of which was rated 'objectionable'!

In view of the fact that organic farming and gardening outlaws all GM products one might wonder why attention should be given to it here. In the first place, if GM crops were to spread more widely there would be untold consequences for both conventional and organic farming based on non-GM seed. Experience shows that genetic contamination is occurring and that the two types of farming cannot ultimately co-exist.[20] It follows that it is no use developing an advocacy of organic and biodynamic agriculture in isolation—part of this advocacy includes a rational rejection of what is a serious threat to biodiversity and human life. It is no use turning a blind eye.

Biodynamic seed development and plant breeding

Following the remarks in Chapter 3 concerning the farm organism, it will be better appreciated why self-produced seed is such an asset for a farm. Moreover, just as the results of optimal nutrition for animals will not fully manifest in the current generation, for beneficial traits to become genetically established in plants it may take several successive years of breeding. So if we are to take on local seed production, what guidelines can be offered?

Firstly, the varieties used will not be hybrids for reasons already mentioned. Open-pollinated and traditional varieties are valuable resources because these can be worked with, and may often be more pest-resistant than hybrids. We should try to find out where these plants have previously been grown, since the biography of each variety imprints an environmental adaptation. Varieties that have been grown under particular climate and soil conditions for many generations are termed 'land races' and are usually considered best for those circumstances, although climatic change will certainly challenge accepted wisdom. Varieties that have 'moved about' and have multiple provenance are better for general purposes and, if subject to seed multiplication, can be distributed with greater confidence. In this respect, commercial hybrid seeds represent a 'one size fits all' approach.

The next main consideration when planning seed production is the type of pollination. *Self-pollinated* plants inherit characteristics only from the parent, so it is possible to have other varieties growing nearby. Lettuce is regarded as a self-pollinating crop as are most cereals. With *cross-*

pollinated plants one should only grow one variety of a particular plant family in the same bed or field, and even wild versions of the plant may cause problems from pollen transfer by wind or insects unless separation distances are strictly followed. In this respect, beans, peas and tomatoes are examples of vegetable plants that are relatively free of problems for seed production. On the other hand, varieties of radish, cabbage, pumpkins/gourds, courgettes, cucumbers, carrots, parsnips and even peppers must be carefully isolated.[21]

The sowing medium must be considered. Seeds will either be sown in the open or into seedling trays, modules or even pots for later planting out. Commercial mixes are available for this purpose but the certified organic grower must ensure that such material is 'approved for organic production'.[22] These mixes will often prove preferable to self-made ones as they not only offer a low-nutrient germination medium of uniform consistency but are free from weed seeds. However, a home-made potting compost can easily be made by taking roughly equal parts of garden soil, compost and gritty sand. This mixture is suitable for development of the rooting system and should be sieved to ensure even consistency. The compost should be well rotted, leaf compost being excellent for this purpose. In this mixture we have a source of organic matter for water retention, and coarser mineral matter to ensure aeration and drainage.

We then come to biodynamic measures. For the biodynamic preparations, horn manure is applied to the seedbed while horn silica is used only after plants have fully established and in the weeks leading up to the seed harvest (Chapter 5). But how should the biodynamic calendar be used?

The essential technique in plant breeding is to identify plants of superior morphology (see below) and then to produce seeds from these. What using the calendar may help to achieve is a higher proportion of plants with the desired morphology from which selection of 'elite' seeds can take place. Sowings for *plant selection* should therefore be made on the appropriate days for the plant organs needing to be expressed (root, leaf etc., Chapter 6). Once seeds from selected plants have been collected we can think about seed multiplication. Here again, the rule must be to continue with lunar-zodiacal timings appropriate for the particular crop plant. Although selection of warmth days will tend to maximize overall seed production and is optimal for fruit or seed crops, choice of such days for root or leaf crops will conflict with the features one wants to select. Since cosmic forces are likely to exert an influence on the character of DNA (Chapter 2), the use of lunar-zodiacal timings provides a legitimate rationale for plant breeding.

Seed selection

As choice of variety is influenced by market needs and such related characteristics as flavour and storage quality, selection of seed for breeding must be made on the basis of 'whole plant characteristics'. These will include healthiness and the way the plant is formed in relation to desired morphology. For example, with lettuce, spinach or cabbage, those which give prolonged leafy growth or good central development should be chosen. Having selected such plants or a block of plants displaying good characteristics, the seed should be allowed to ripen well on the plants before collecting (Figs 7.1–7.2). In some cases it will be necessary to support stems or to place a bag over seed heads to avoid seeds scattering on the ground. Several rows of seed plants should be sown together so that selection can take place from *inner rows* which experience more protec-

Fig. 7.1 Lettuce, leek and onion seed crops in a polytunnel at Bingenheim, Germany (Peter Brinch)

Fig. 7.2 Carrot seed crop at Dottenfelderhof, Germany (Peter Brinch)

tion. It is from here that the best plants are usually found, and from these, elite seed for subsequent improvement of quality is taken. Care must be given to the labelling of rows and to identification of superior plants.

The selection procedure adopted by breeders thus identifies *elite* seed, from which plant breeding will continue, and *standard* seed, which is the quality sold to the market. In reality today's elite seed will very likely become tomorrow's standard seed following its multiplication. Elite seed results from rigorous selection. It may only be collected from 10–15% of the plants—positive selection. On the other hand, negative selection may be adopted. Here, up to 20% of plants which exhibit weakness, disease, poor shape or which go to seed too early are 'rogued' (pulled out or cut) to leave a remaining superior crop for seed production.

While elite seed will normally be harvested from chosen whole plants, more sophisticated selection may be carried out. For further selection and research this may involve taking the first seed heads to develop (known as the 'King's Head') or selecting seed according to its position on the stalk—cereal grains for example. The higher germination rate of such seed points to its superior quality and vitality. Elite seed is often subject to what is known as progeny testing to ensure that desired traits emerge consistently in subsequent generations.[23]

Biodynamic plant breeding has developed mainly in Germany, and in recent years has taken a new direction under the guidance of Dorian Schmidt. This approach is based on developing faculties of observation in the field of nature's formative forces. Trained observation of subtle energies in the plant kingdom has helped to bring an understanding of otherwise hidden attributes present in our food plants. These show themselves to the trained eye as colours, as lightness or darkness, and as movement and gestures. The qualities arising from such observations are used to determine whether plants carry suitable inner qualities of nutritional value to be used for further breeding. In this way, a number of food plants have been identified as having qualities of healing, whereas certain varieties—mostly modern ones—have been experienced as unsuitable for human nutrition. Perceptual research does not, of course, rule out the need for suitable physically based breeding methods to be used, but in aiming away from a purely materialistic direction it offers a new basis for selection of future food plants.[24]

Post-harvest procedures

After seeds have been collected they will require to be thoroughly dried (Fig. 7.3). Seeds from moist fruits such as cucumbers and tomatoes need to

Fig. 7.3 Seed crops drying in a barn at Stormy Hall, UK (Peter Brinch)

be fermented in water to remove gelatinous coating. They are then washed and dried, preferably in the sun, without delay. After drying seeds, cleaning and grading takes place according to size or other characteristics. This ranges from the use of sieves and sorting on trays, to winnowing by the wind or a fan, or the use of special machines (Fig. 7.4). After grading, seeds should be stored unharmed until required for sowing. Adequate space must be allocated for this, and the professional seed producer will often use a cupboard with multiple shelves kept at constant temperature and humidity prior to packing for sale.

Seeds have variable longevity with most vegetable seeds lasting at least two years under temperate conditions. But since the survival of seeds is affected by conditions of storage, and some have limited viability, seeds of important varieties should be saved each season.[25] Hot and humid environments limit the time that seeds can be safely kept. However, if

Fig. 7.4 Seed cleaning machine at Stormy Hall, UK (Peter Brinch)

clean seeds are stored in a protected dry environment, in airtight jars or well-sealed polythene bags, they should escape attack by insects such as weevils. In cases where there is a risk of insect eggs already being inside the grain, there is little that can be done except to introduce carbon dioxide or nitrogen into the storage vessels before sealing, but this is usually beyond the capacity of the individual grower. French beans and broad beans produced by organic growers are routinely deep frozen to kill insects. In cases of severe pest infestation one might resort to heat treatment.[26] Sometimes plants grown for seed are themselves affected by disease. In this case, a precautionary treatment before sowing would be to place the seed in hot water at 50°C for 30 minutes. Such a method is very effective but is a more practical proposition for vegetable seeds than for cereals.

Before sowing on a field scale, and especially before sale or exchange, it is essential to test for viability by determining the *percentage germination.* This will be more important the longer that seeds have been in storage. It may take two weeks or more, depending on the type of seed. Having freshly mixed the seeds, count out 20 or 50, according to size, onto a piece of moist absorbent paper. Some seeds require to be kept in the dark but many are better placed in light and warm conditions. After a time, count the number of seeds that have germinated and calculate the percentage.

The process of seed germination

To begin with, water is absorbed. The result is that an internal process takes place with production of the auxins—cytokinin and gibberellin—together with ethylene gas, all of which set about transforming the store of fats and oils to fatty acids, and subsequently to sugars and starch. An enlargement is noted as cell division starts to take place, causing also a secondary phase of water uptake. Only after this will the first root, the radicle, and later the first upward shoot, the plumule, emerge. During this stage the food reserves of the seed are being 'digested' and translocated, a process that continues until the plumule's cotyledon breaks the soil surface and photosynthesis commences.

Water is the crucial link between this 'scientific' explanation and a more metaphysical one. As we shall see in Chapter 8, water is the great catalyst for the working of life forces. When seeds are placed in moist soil they undergo an awakening process. The seed with its cosmic imprint unites itself with the earth and the life forces it finds there, a fact reflected by the radicle being the first vegetative growth to emerge.[27] It experiences the working of cosmic ethers, in essence the same as a human being's relationship to the constellations at the time of birth. This is the ultimate basis for the use of a biodynamic calendar.

Germination efficiency and priming

Slow and uneven germination remains a widespread problem in world agriculture. This means that the young crop does not make best use of soil and water during crucial rains or irrigation. Slow germination also leads to loss of seed to predation, pest attack and fungus disease. Poor early development has an adverse effect on later yield, and as farmers and growers will be paying for at least some of their seed the consequences of low yield are clear. The failure of crops to develop quickly and uniformly depends on a number of factors, the most obvious of which is seedbed quality—in particular, the lack of proper contact between seed, soil moisture and air.[28] Choice of variety and seed selection also influence the outcome.

Seed drilling as opposed to broadcasting, should improve germination efficiency, and in theory use and waste less seed. But this requires the equipment, owned or hired, to carry it out—certainly not possible for those in poor countries. Although predation of seed can be reduced by pelletizing—encasing it within a larger granule—this again is only warranted in markets which support relatively high value products. In organic and biodynamic growing, pelletizing is usually done for ease of handling and accurate spacing of seed. In conventional, chemically based farming, whether pelletized or not, seed is usually treated to protect it from attack by fungus.[29] So in what further way might the farmer or grower improve their seed germination rates? In different parts of the world measures are taken for preparing seeds before sowing in the field or nursery so that germination is accelerated. Such methods, including freezing and a variety of wetting treatments, are collectively known as *priming*.

The wetting procedure depends on the type of seed, and how it can be handled and surface-dried before sowing. As a general rule, soaking overnight is a sufficient period for all but the largest seeds to absorb a significant amount of water, enabling them to germinate more rapidly on contact with soil. Seeds all have their 'safe limits' beyond which soaking may cause damage, while sowing of soaked seed should not be delayed beyond 12–24 hours. Awareness of the full moon period as a time when water uptake is more rapid will be helpful.[30]

Experience with seed-wetting methods shows they incur little risk or cost, and have been effective in raising yields. The reader must judge whether for them, time involvement or other issues are likely to be a deterrent. In the writer's experience, a more serious issue is a lack of awareness of the opportunities these methods offer. Commercially available seeds may have already undergone treatment to assist germination so it would be good to check this.

Biodynamic seed treatments

Biodynamic preparations are sometimes added to the water used to prime seeds. Such *seed-baths* or *seed-soaks* combine the standard benefit of water absorption with the stimulus afforded by the preparations.[31] It has been shown that 'treating seeds in this way increases root formation and produces plants that are more disease-resistant and generally much stronger'.[32] So in stimulating plant growth over a longer period than the germination phase this biodynamic procedure is not simply a priming technique (see panel below). How then, do we visualize these materials acting on the seed with such beneficial results? The simplest explanation is to say that the preparation raises the etheric or life quality of the seed. This in turn stimulates root development and acts to deter pathogenic organisms. As if to underline this principle, cow's urine is used around the world for repelling insects and pathogenic organisms. Like the preparations, this offers strong life forces.

Suggestions have been made for using various preparations: *horn manure* for all crops, especially roots and spinach; *chamomile* for peas, clovers, beans, radish, brassicas and linseed; *oak bark* for oats, lettuce, potatoes and beans; *valerian*—connected with a warmth principle—for germination of most cultivated plants including wheat, fodder beet, sugar beet, carrots,

The effects of treating seeds with very dilute solutions (0.003%) of biodynamic preparations (after Pfeiffer, 1983)

A. Radishes (grams, average of 20 plants)	Radish and Root	Leaves
Control without preparations	2570	2530
Seed bath with 500	2775	2655
Seed bath with dilution of BD compost	3010	2300
Seed bath with 500 plus 502–507	4600	4090

B. Maize (average for 10 plants)	Height (cm)	Weight (g)		
		Roots	Stems	Ears
Control without preparations	181	50	2181	727
Seed bath with 500 plus 502–507	185	65	2250	800

chicory, cucumbers, tomatoes, peppers, pumpkin, onions, leeks, celeriac, celery and potatoes (it should also help promote germination of legumes under cool conditions).

The method in general use involves taking one teaspoon (1–2 cm^3) of each solid preparation, stirring into one litre of rainwater and leaving to stand for 24 hours. Seeds are then put in a cloth bag and immersed in the liquid for 10–15 minutes. They are then spread out to dry, or sown immediately. In the case of valerian, the expressed juice from valerian flowers is diluted by adding one tablespoonful to 10 litres of rainwater and stirring for 10 minutes. Seeds can then either be suspended in the valerian preparation or sprayed with it, but no more than a day before sowing. In the case of harder seeds which take longer to germinate, longer soak times may be beneficial (but see below).

For larger quantities of seed, as with cereal grain, one tablespoon of preparation is added to half a bucket of rainwater and left for 24 hours. This liquid can then be diluted if necessary. Approximately 3 litres of liquid is needed for 10 kg of seed. The seed is emptied onto a clean floor or other surface—perhaps a large sheet—then sprayed, turning the seed to ensure it all gets properly wetted. The seed is then covered with sacks for 12–24 hours (preferably the shorter time) before sowing, being careful to protect the seed from predators such as rats and mice. The seed can now be allowed to surface-dry before sowing. Very small seeds are normally sprayed with preparation and allowed to dry before sowing.

If the above preparations are not available for seed treatment, 'cow pat pit' preparation or even compost extracts might equally well be used. Cow's urine was mentioned above and can be diluted 50:50. There are now also potentized preparations available for this purpose.[33] It should always be remembered that whatever creativity lives in the mind of the gardener or farmer—such as the use of healing herbs—has a force recognizable to the realms of nature.

Biodynamic wetting times are shorter than for standard seed priming. Here, our comments on the process of germination should be recalled. If seed is given a prolonged wetting time and undergoes its awakening period when it is not yet in soil, then it cannot fully benefit from influences being transmitted to it *via the soil*. On this interpretation, if the biodynamic calendar is to be used, priming should be brief and should not precede by more than a few hours the appropriate sowing/planting sign indicated in the calendar.

Following the above principles, several steps are recommended. First, make a choice of the priming material and priming method to be used for the seeds in question. Then, if you are going to use the biodynamic calendar for sowing, find the appropriate times for the particular plant. Prepare the seedbed, pots, trays or modules. Finally, carry out both the priming procedure and the sowing early in the appropriate lunar-zodiacal period.

Biodynamic root treatment for transplants

Just as seeds can benefit from the treatments described here, biodynamic practitioners have used solutions or slurries, often referred to as *root dips*, with which to treat transplants. While various recipes might be advocated, the most obviously beneficial foundation will be horn manure with or without 'cow pat pit' (Chapter 5).

The main value of root treatment will be for transplanting shrubs and trees, and in such cases a slurry can be made with some clay soil or bentonite which adheres to the whole root system. Otherwise, treatment of root systems must be done according to the risk of damage. When re-potting or potting-on, soil in the immediate vicinity of a transplant might be treated if there was serious risk of physical breakage to an exposed root system. In the case of all seedlings such as annual salad leaves, individual root treatment is unrealistic as well as being risky. In this case the procedure should simply be to apply horn manure with or without CPP while still in the modules or trays, or alternatively, immediately after planting out.

8. Water: The Foundation of Life

It seems too obvious to say that water is essential to life. Most living organisms, including ourselves, are predominantly water. In this chapter we shall reveal some of the less recognized characteristics of water and concentrate on aspects that have a particular relationship to biodynamics. This will require that we are prepared to see water not only as a carrier of substances but also as the mediator of life forces.[1]

The special significance of water

Thales of Miletus (640–546 BC) stated that water was 'the only true element, the original substance of the cosmos, and imbued with the quality of Being'. 'Everything originates from water, and into water all things return again.' In water, hydrogen and oxygen are combined. We earlier characterized hydrogen as the carrier of the primordial spirit and oxygen as the carrier of the life force. It seems that implicit in the knowledge of classical Greece was that 'life' originated as spirit, a condition to which it eventually returned. Viktor Schauberger regarded water as 'the manifestation of sublime forces—the offspring or first born of these cosmic energies', and that it was 'a living substance'.[2] A sense of the latter is conveyed if we think of the cycle of water extending from the bowels of the earth to the heights of the stratosphere.[3] Rudolf Steiner stated that 'the astral atmosphere of the earth lives in the circulation of the water'.

Is it any wonder that from the Bronze Age into Celtic or Roman Britain the occurrence of water was marked by countless sacred springs and holy wells. Water has always had a special place in ritual—we can think of how it is used to confer blessings in the Christian church and traditional Buddhist *pirith* ceremonies. In Sanskrit, a sacred written language, there are some 30 different words for water.

Like orthodox science, biodynamics considers living organisms and water as indissolubly linked but the emphasis is different, for here we are considering life forces to be primary and matter secondary. In this conception, water's role becomes that of mediator between life forces coming from the periphery, and the physical organisms themselves.

Though carbon is essential to the manifestation of life, water is that earthly medium without which other processes cannot occur. We might say it acts more like a catalyst than a primary player in the manifestation of life. We perceive few changes to water in the process—it is colourless, transparent, lacking in fragrance, formless and itself apparently lifeless. But such lack of assertiveness is the entry point for our appreciation of water. Water takes nothing for itself while so much happens around and because of it. In terms of physical processes, water conveys dissolved gases and nutrients while it takes away the products of metabolism needing to be eliminated. Between death by too little (drought) or too much (saturation, drowning), a sufficiency of water enables life to come forth in its phases.

Important physical and chemical characteristics of water

Water has a number of unusual properties. In the first place, for its density it has a particularly low viscosity, allowing—even at low temperatures—rapid rates of random movement (Brownian motion) of contained substances and micro-organisms. Such a characteristic is conducive for life processes. Then, for a compound of low molecular weight it has a surprisingly high boiling point. That water exists as a liquid over such a wide temperature range is due to innumerable hydrogen bonds which hold individual units of the structure in place.

We should note the configuration of the basic H_2O unit. Its angular symmetry leads to wide exposure of a negative pole in the molecule (Fig. 8.1). This attracts hydrogen ions which then proceed to link adjacent molecules. There is thus a tendency for systematic linking of molecules, or molecular clusters, through the entire liquid. It has been argued that life could not exist without such coherence. Such pseudo-crystalline microstructure appears to take different forms according to temperature.[4] Plato was apparently able to see water as consisting of icosahedra!

As the solid state is approached, density increases, reaching a maximum at 4°C. The formation of ice crystals is accompanied by a 9% *increase* of

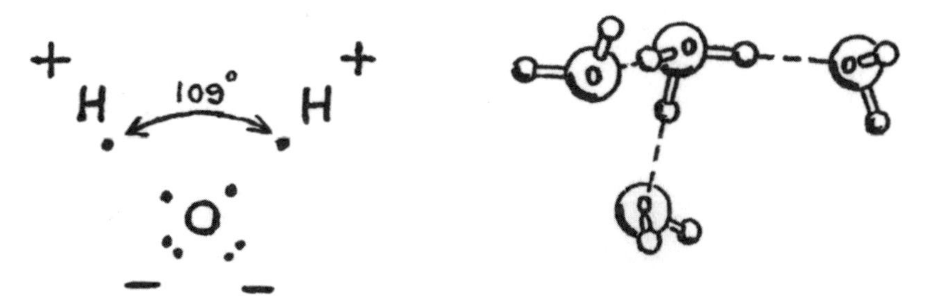

Fig. 8.1 The water molecule. Left: The water dipole. Dots indicate electrons contributed by each atom. Right: Complex ionized unit building to form three-dimensional structures. Dashed lines are hydrogen bonds

volume compared with the equivalent amount of liquid. Ice therefore floats on a body of near-freezing water, so protecting life below. Frost commonly forms on the surface of leaves when the atmosphere reaches 0°C. Frost damage occurs when ice formation ruptures biological tissues, but due to substances dissolved in cell water this will generally not occur until temperatures are several degrees below the freezing point of pure water.

At higher temperatures, water possesses simpler structures as reflected in its degree of ionization—the extent to which H^+ and OH^- ions are present. At 10°C the so-called ionic product is around 0.3 moles/litre^{-1}, at 22°C this has risen to 1.0 and at 40° it is 2.9. The higher the ionization, the more active water becomes in its chemical interactions.[5] This is the basis for the rule that for every 10°C rise of temperature, rates of chemical processes double. In the lower range, the same applies to biochemical processes but upper limits for biological processes are set by other factors. As temperature increases, less gas can be dissolved in water, meaning that metabolic functions based on oxidation-reduction reactions are retarded or prevented.

Besides ionization itself, we should consider the *balance* between water's positive and negative ions. Where hydrogen ions are in excess we refer to acidity, where hydroxyls predominate, alkalinity. The well-known pH scale thus refers to changes in the balance between H^+ and OH^-.[6] There are always polarities at work in living organisms and here we have a prime example. Life processes respond to changes of pH—our digestive process, for example, presiding over some remarkable contrasts. An important rock weathering process known as hydrolysis is dependent upon the supply of H^+ ions. These are generated, for instance, by the reaction of CO_2 and organic acids with water. For all the above reasons, under acid conditions, as temperature is increased so will be the rates of rock breakdown.

Water's characteristic of generating transient ions and pseudo-crystalline formations (clathrate or cagelike structures) allows it to carry much dissolved and suspended substance. So, even from a mainstream point of view, water is intriguing and poses mysteries. This has meant that many who are interested in water are willing to consider investigations that depart some way from standard physical and chemical analysis.

Water and homoeopathy

Among the studies most relevant to biodynamics are those of Theodor Schwenk, Lili Kolisko and collaborators.[7] Schwenk's inspired publication

Sensitive Chaos has been an indispensable primer to understanding water's ability to mediate life processes—to be the carrier of pattern and rhythm. On the other hand one might argue that the pioneer contribution of Samuel Hahnemann should be placed first.[8] His insight into the nature of substance, together with that of Rudolf Steiner and, later, Rudolf Hauschka, undoubtedly forms a basis for our present discussion. However, an account of homoeopathy as normally understood would be too much of a digression at this point. Suffice to say that fundamental to the *organon* is that water can take on the specific quality of a substance, in other words, memorize the 'patterns' of other substances.[9]

In his research, Schwenk built upon the work of Lili Kolisko.[10] Since organisms manifest life through the medium of water, he enquired whether it is possible to show that the potentizing medium is responsive to formative forces when surfaces are created within it. In a notable experiment, his team prepared 12 flasks of distilled water. At two-hourly intervals over the course of the 24-hour cycle they shook one flask in a steady rhythm for four minutes. The water was then used to germinate 50 wheat seeds which were monitored as they grew to approximately 10 cm. The same procedure was applied using unshaken water. Sprout lengths were plotted against the time of day at which the shaking took place. This was repeated over a period in order to establish any characteristic pattern. The shaken water gave rise to rhythmic variability while none was observed in the unshaken controls. Seedling performance typically deviated 5% above and below a mean and provided a background against which to measure anomalies. The latter were noted around an eclipse—the water shaken during this period produced plants with poor growth.

A similar experiment was conducted by Enzo Nastati and his team.[11] Germination rates of wheat seeds were plotted against the time at which the water had been shaken *three months previously*. The split experiment had one lot of seedlings grown in water alone while the others were given water containing potentized gold. The former seedlings were subject to a reduction in germination at an eclipse, the latter were not. This is explained by acknowledging gold, as did the alchemists, as the earthly representative of the sun. Providing it in homoeopathic form compensated for lack of physical sunlight. In Chapter 5, we encountered a similar principle when discussing use of the horn silica preparation while Pfeiffer showed that biodynamic preparations could compensate for lack of lime.

An experiment of Rudolf Hauschka adds to this picture. This measured the amounts of CO_2 produced by the growth of yeast—cultured with solutions of either natural or synthetic benzoic acid.[12] For this purpose, the two types of benzoic acid and a water 'control' were subject to serial potentization beforehand. From the amounts of gas produced, it was evident that with benzoic acid of natural origin, variations in the life activity of the yeast were measurable with progressive potentization, whereas a constant low-level response was elicited from the synthetic chemical. As expected, no variations were observed from the water control.

Let us now try to summarize this picture. Water appears to be sensitive to external forces. These it can accumulate by certain shaking methods (and see below vortex and flowform). It appears able to retain these forces for periods of time. The retained forces influence the growth of plants and are therefore connected with life. Added substances in potentized form also release their qualities provided they are natural rather than artificially synthesized. As growth responses vary with changing potency, it shows that vital forces are liberated according to rhythmic laws.

A certain parallel may be drawn from plant nutrition. Cadmium, mercury, lead, arsenic and chromium at extremely low concentrations stimulate the growth of seedlings, while as concentrations rise growth is inhibited. Also low concentrations applied around germination protect experimental plants from adverse effects of higher doses applied later. The fact that low concentrations have a growth response is certainly reminiscent of homoeopathy where potentization (low concentration) is the normal cure, while high concentration is associated with pathological symptoms. In such cases it appears that 'radiations' are tolerated or even beneficial, while too much chemical presence is not.

If we return to the molecular characteristics of water we can easily appreciate that it is just such an ionically active and dynamic structure which has the capacity to convey subtle forces. Furthermore, once a certain 'resonance' is set up in the liquid, physical dilution cannot erase it.[13] While offering insight into homoeopathy—even to the action of the biodynamic preparations—such knowledge adds a new dimension to our understanding of pollution.

Water quality and remediation

Natural water is not pure water! Pure water, as we have characterized it, only exists as distilled water. Less argumentatively perhaps, it occurs as rainwater—but even then, because it contains dissolved CO_2 it will not normally exhibit neutrality (pH 7). And the bulk of precipitation requires microscopic nuclei around which to condense, so trace amounts of dust are nearly always present. Even the 'pure' water in a mountain stream contains dissolved substances from the rocks in its catchment. Typically, as you follow a stream from its source to the sea, the amount of dissolved substance (measured by electrical conductivity) increases. These substances are joined by many others which enter the river from organic sources as well as from human actions. Still further materials, from soil erosion for instance, are held in suspension, creating turbidity.

Human beings actually need the minerals that water contains—and it bears many substances that are not necessarily problematic. For this reason we can regard a wide range of different waters as being equally 'good'. Many of us depend on ground water for our supplies and this will normally contain more dissolved substances than surface waters. Spa waters, containing characteristic assemblages of minerals, have traditionally been regarded as efficacious, so we should certainly not assume that the best water is as close to pure H_2O as possible. On the other hand it is clear that too much of almost any substance limits its usefulness and, taken on a regular basis, may eventually endanger health.

There is then the question of water purification. In many countries, water in the public supply has been through a purification process. The Austrian saying that 'seven stones clean water' refers to nature's own cleansing process as stream water passes over rocks. In modern water treatment works, air is forced into water so that a digestive process can take place while it passes through a succession of filtration beds. The whole process ensures maximum permitted levels of mineral substances and pathogenic organisms—in all, more than 50 different parameters may be monitored. Standards have been set below levels which are believed to affect health but conventional wisdom cannot take account of long-term impact nor of individual susceptibilities. For example the limit set for surfactants is 200 $\mu g\ lit^{-1}$, copper and zinc 3000 and 5000 $\mu g\ lit^{-1}$ respectively, and pesticides 0.5 $\mu g\ lit^{-1}$. Assurances such as these may be of little consolation to allergy sufferers, but on the whole we should be grateful that in such a highly populated country as Britain there are controls in place on the quality of potable water. On the other hand, the issue of introducing fluoride into the water supply has ignited public concern no less than GMOs (Chapter 10).

In recent years there has been a move to integrate nature's way into the treatment of sewage. Water has been allowed to pass through plants in constructed wetlands—reed bed and pond systems—while flowforms (see below) have provided a means both of aeration and of re-establishing natural rhythm into water[14] (Figs 8.2 and 8.3). Given a suitable fall in land level, the reed bed or pond system is able to address requirements up to the level of small communities, offering an alternative to septic tank

Fig. 8.2 Reed bed system at Oaklands Park Camphill Community, Gloucestershire (Mark Moodie)

Fig. 8.3 Close-up of flowform cascade in Fig. 8.2 (Mark Moodie)

seepage.[15] Beyond this, the volumes of sewage from towns would need proportionally larger areas for processing, land which is simply not available.

Agricultural water quality

We should be conscious, too, about the quality of water used in agriculture. This applies in irrigation, and especially in covered horticulture. Here, accumulation of salts in the soil becomes possible in the course of time due to strong heating and an upward capillary process in the soil. This is a particular concern if irrigation water comes from underground sources and has a high electrical conductivity. Commonly this affects soil and plant growth by raising the pH due to lime content. It is a problem more likely to occur if polytunnels are not annually dug over and fresh compost introduced, for as organic matter diminishes, soil structure is lost and capillarity increases. In non-temperate regions impure irrigation water also leads to soil salinity, which restricts and eventually prevents plant growth due to the raised osmotic value of the soil.

Rainwater is superior to river or groundwater, while mains water (the piped supply) will often be the least desirable owing to its chlorine or lime content. After all, the response of grass growth to a thunderstorm is impressive compared with sprinkler irrigation! If there is no choice but to use treated water from the public supply—for example when applying preparations or watering seeds—it should be left to stand for several hours or overnight. This will eliminate most of the chlorine which otherwise has an adverse, if temporary, impact on micro-organisms. It is this type of water, together with groundwater or any grey water for reuse, which will benefit most from a rejuvenation or vitalization method as described below. If at all possible, install a system to harvest rainwater for indoor cropping.

Non-standard water quality evaluation

Aside from standard chemical and microbial analysis, a qualitative method known as *sensitive crystallization* (Chapter 10) is relevant to an overall evaluation of water. This involves taking a sample of water and placing it in a standard solution of a salt such as copper chloride, which is then very slowly dried in a Petri dish at around 30°C. The type of crystal formation that results, when compared with samples from other locations—as well as samples taken at the same location on different occasions—is claimed to characterize the water's life forces. This is a demonstration-pictorial method, for it can only rank samples according to the formative gestures of the crystals. Similarly, Masaru Emoto has found that the way water forms crystals when it freezes bears a relationship to its purity.[16]

A further approach, known as the *drop-picture* method, has been devised by Theodor Schwenk.[17] This requires drops of distilled water every five seconds into the centre of a dish of the test water sample thickened with glycerine. Currents radiate from the centre to the edge, returning to form an interference pattern. This is repeatedly photographed and in this way it has been shown that characteristic forms appear after particular numbers of drops. These forms reflect the relative mobility of the water (Fig. 8.4). Like the crystallization and chromatographic methods, this is a qualitative approach but, even so, waters judged pure by normal methods have been found to have very different 'signatures' according to their previous history.

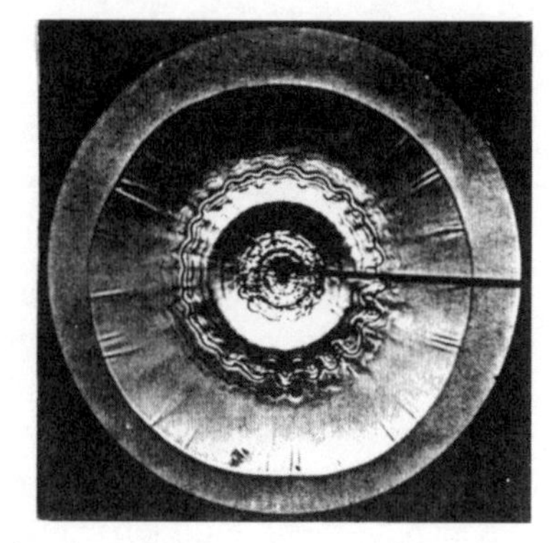

Fig. 8.4 Comparison water drop pictures. Left: Water from a stream near its source, Schwarzwald, Germany. Right: Polluted water from same area, downstream. (From Theodor and Wolfram Schwenk, 1989)

Vitalization of water

Water's apparent ability to record external stimuli means that even when it has been treated and is deemed fit for public consumption it can still carry an unhealthy quality. Water from a pure source that has been bottled may be 'damaged' by processing prior to bottling, while use of plastic bottles should be regarded with suspicion for the reason mentioned in Chapter 10. While this presents a negative picture of processed water, the concern here is simply to show that conventional water treatment leaves certain questions unanswered.

Water is sensitive to vibrations—vibrations in a physical sense, also in an electromagnetic and molecular sense and, as we discovered with potentization research, even in a 'cosmic' sense. How then can we 'heal' water as opposed to purify it? If water's character can be degraded by combinations of substances and energies, with suitable knowledge it ought to be able to be restored to a satisfactory condition. Certain lesser-known techniques thus aim towards the remediation of water. These 'iron out its creases' and re-establish a more natural pattern. Water responds beneficially to vocal sound as in the example of 'Tonsingen' just as it does to traditional Eastern chanting. Water may be led down pipes with special internal surfaces, stored or passed through pottery vessels of particular characteristics or subjected to systems of filters. In some of these, the beneficial vibrations of crystalline silica play a key role.[18] In biodynamics, the vortex and the flowform are used for water activation.

The vortex

Spirals and vortices are forms which appear on the largest (galactic) and smallest (atomic) scales, with cyclonic weather patterns being intermediate examples. Thus was Charles Leadbeater, as a result of intense meditative practice, able to describe what was referred to as 'the ultimate physical atom'.[19] Here, 'ultimate particles' are seen as whorls of energy manifesting in positive and negative spins, the negative version being shown in Fig. 8.5. Well before modern nuclear physics, this portrayal has been found to be in remarkable accord with 'super-string theory'.

Vortical motion is a natural behavioural characteristic of water in a river when parts of the flow proceed at different rates. The form of the vortex is a spiral when viewed from above (Fig. 5.4) and a parabola from the side (Fig. 8.6).[20] The water in a vortex forms many thin layers which slip past each other, and water drawn in at the centre re-emerges at the periphery. Through the vortex, water can receive new information patterns. Percussive shaking, as in homoeopathy, generates turbulence which incorporates countless micro-vortices allowing the vibrational pattern of an added substance to resonate in the water.

The spiral vortex form appears to have a special relation to force fields (ethers) coming towards us from the stars and planetary worlds. The ancient Celtic people regarded the right-hand (clockwise) spiral as representing dematerialization while the left-hand (anticlockwise) spiral

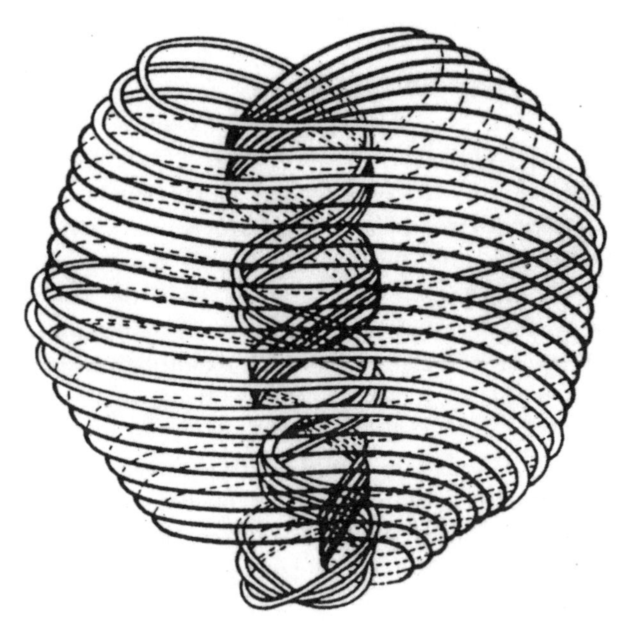

Fig. 8.5 The 'ultimate physical atom' (from Leadbeater and Besant, 1951)

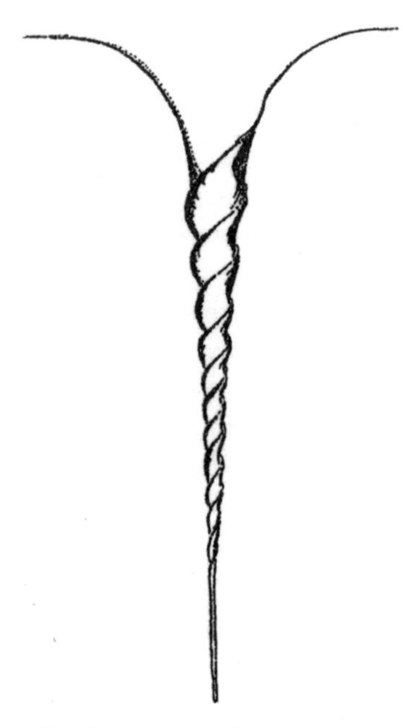

Fig. 8.6 Sectional view of vortex

represented materialization or the entry of spiritual forces into matter or substance. In Fig. 8.7 we see this ancient Celtic spirituality transferred to that of Christ. Dematerialization was associated with daytime and summer, materialization with night-time and winter. Witches and other alchemists used full moon nights for their materialization processes, as full moon increases the densification of matter, new moon the reverse.[21]

This offers insight into how water may be restored by vortex treatment and why the biodynamic preparations were first stirred in this way (Chapter 5). But there is a further aspect. Between alternate vortices a chaotic pattern is generated in the water. Steiner commented that whenever chaos is present, cosmic formative influences can enter the physical world—a principle we encountered when discussing seeds (Chapter 7). There is thus a complementary relationship between vortex and chaos, expressed in the thought that 'many tangent planes mediating cosmic forces are caught and held at these moments of chaos'.[22]

Vortex treatment is again graphically expressed by Enzo Nastati who writes: 'Such thin inner surfaces, if placed on a single plane, would cover incredibly large areas. One can calculate that in a litre of water such surfaces would be able to cover 400 hectares. Such a surface can be compared to an enormous parabola, able to receive the greatest amount of forces from the cosmos. Such an enormous dynamic and sensitive surface collapses into the volume of water when the vortex suddenly ends, provoking an implosive effect that traps the cosmic forces picked up in the phase of expansion. All this would be useless if the water were not then able to retain the imprinted message (resonance) in its "memory"'.[23]

Regularly reversing the rotation of the vortex would therefore seem to invite a cosmic breathing process to occur—a process likely to restore the natural qualities and vibrancy of water, improving its capacity to support life. Such water has added energy to empower the biodynamic preparations.

Fig. 8.7 Celtic spirals on eighth-century Irish bronze motif (National Museum of Ireland, Dublin)

Flowforms

Flowforms are vessels whose particular shapes engender rhythm in water flowing through them. The principle was discovered by John Wilkes in 1970 while he was engaged in a research programme initiated by George Adams and Theodor Schwenk. The problem was to find a way in which water could be made more sensitive to etheric formative forces. Research was conducted into the forms and processes of the natural world—a world in which water always participates. Particular attention was directed towards the ways in which water naturally moves. Wilkes assisted by sculpting mathematical surfaces corresponding to the geometry of these forms of the living world.[24] By experimenting with symmetrical shapes through which water could flow, rhythm arose. Subsequent development of different designs and applications has been due to Wilkes and collaborators.

Flowforms, then, comprise bowl-shaped surfaces through which the water flows; they may be single units or rosette arrangements but usually are set up in multiple cascades (Figs 5.7 and 8.3). There is a great variety of movement in the different designs, many generating a lemniscate or figure-of-eight pattern while, more recently, forms have been made that produce a large-scale movement in alternating left- and right-handed vortices.

Flowforms are now well known as artistic water features and have been widely used to stir biodynamic preparations and the water for baking and fermentation. They have also found application in pond aeration for reed bed systems and fish farming, as well as for treating irrigation water. Flowforms are widely used to treat slurry, in this way processing large volumes of liquid in a highly practical way. While more research needs to be conducted, flowform-treated water has been shown to have positive effects on germination, plant rooting, crystal formation and bread quality, showing that the rhythmic treatment of water affects biological processes.[25]

Final thoughts

For water, we can think of its movement in outer nature as a kind of freedom while its participation in form creation and metamorphosis is a selfless act in the support of life processes. This, after all, is the substance that carries a universal consciousness (hydrogen) together with the earth's life forces (oxygen), so its menial tasks in the service of the world make it a worthy subject for meditation. As colour manifests between darkness and light, the rainbow surely reveals a further manifestation of the secrets of water. Many today will say that all this is 'living in dreamland', which sadly bears out Thomas Carlisle's statement, 'The progress of science is to destroy wonder.'

So despite growing knowledge about water and its importance to human life, science and industry do not want to be confronted with ideas indicating there is more to water than its chemical properties. Instead, it is left to individual commercial ventures to tout their eccentric devices in the marketplace while the general public, educated by an unsympathetic and fearful system, remain in a dream-state about the true nature of the fluid that controls their lives.

Consider what happened to Jacques Benveniste some years ago when 'science' confronted his contribution to allergy research. The root cause of the rejection of homoeopathy by orthodox science is that a physical causation must always be found, whereas none will occur when only an intangible force is involved. But this work, although published in the journal *Nature*, was perceived as far too damaging for acceptance by establishment science.[26] Such has been the fate of Schauberger and others besides. We have here a real battlefront between the 'old guard' of science, anchored largely by professional and commercial interests, and those whose ideas could contribute so much to the future.

9. Healing Outer and Inner Landscapes

By Margaret Colquhoun

In Chapter 1 a particular point was made about our regression from earlier times when, so far as we are able to tell, a better sense of balance existed between human beings and their environment. The need to strive for attainment of balance in our times is a consequence of having fallen out of balance at some stage, for our outer landscapes are largely a projection into nature of what we bear within ourselves. This applies to urban planning and architecture as much as to rural landscapes. Harmony or disharmony within generates its counterpart without—and the converse.

Although rarely conscious of this, we long for places where the world is still intact—where our souls can breathe in a more ordered if not natural space. An example still familiar across Europe is the village with church as its focus, surrounded by largely medieval patterns of fields, with common grazing and woodland beyond. Many such places have become 'landscape museums' for tourists—places we stay at or visit on holiday because they are quaint, traditional, and remind us of what we have lost. They make us feel whole for a while. Greek islands, high mountain villages, ancient city centres, all have the same effect.

But going on holiday would formerly have been a strange concept. In the past a holiday, or holy day, was a moment when people stopped what they were doing in daily life, washed, dressed nicely, enjoyed and celebrated the special nature of the time and place. There was a feeling that this landscape 'is where, and who, I am and you are—let's enjoy it, improve it, celebrate it'. By contrast, in this modern age we have turned many of our outer landscapes into prairies of boredom, monotonies of monoculture. We've lost sight of the connection to the stars, to sunrise and moonset. The pollution of street lighting blights our vast urban and suburban landscapes and cuts us off further from our connection to the universe. Some of us know neither the names nor how to recognize the constellations—the same for our native trees! If our landscapes outside resemble deserts, what has happened to the landscapes of the soul? Where is the biodiversity, harmony and healing of the 'gardens within' which might obviate the need for holidays in foreign climes. Where are the erstwhile shrines and altars of our fields and hedgerows? What has happened to the sacred places of our hearts, the inner shrine of the soul? Can I buy these on Ebay or at the supermarket? Can I find them on a DVD or through the internet?

Setting the biodynamic scene

In this chapter, we will be concerned with the recreation of harmony in both inner and outer landscapes and will gain insight into how biodynamic agriculture, when consequently carried out, has the potential for healing ourselves and our landscapes.

In his agricultural lectures,[1] Rudolf Steiner emphasized the importance of considering an agricultural unit as an organism. We are urged to treat a farm in such a way as to achieve balance just as our bodily organs are functioning in a harmonious way when we are healthy. As a whole, these lectures lead us on a journey of self-development towards healing our farms. Karl König, the founder of the Camphill movement and a practising medical doctor, gave many lectures to biodynamic farmers and gardeners in the 1960s. In one of these he said:

> I have given much thought to what Rudolf Steiner gave to the world when he gave us the knowledge of these preparations [*the biodynamic spray and compost preparations*]. He has shown us the way to make the soil healthy, and thereby to improve the health of animals and men. These preparations bring about balance in the life of the earth. Therefore it would be impossible to take too earnestly what the spiritual leader of our epoch has given us. The balancing forces for the whole life of all the kingdoms of nature have been handed over again to mankind. This is an event corresponding to what happened in the Persian Epoch, when grain was developed out of grass. I am convinced that what we received at Koberwitz is of the same importance. We must use the methods Rudolf Steiner indicated with great exactitude and with an ever-increasing attention to the details. Without this, the balance of the whole household of nature would soon come to an end.[2]

What did König mean by healing forces for 'the whole life of all the kingdoms of nature'? This is mighty. If we need healing forces, does this mean we are unwell? Is the earth sick? Are we sick—and the animals and plants? What does it mean to be sick or unwell? Is it something like being un-whole? The modern expressions holistic or wholistic seem to be pointing to a new way of viewing and doing things—a healing, more wholesome or holy way.

How did we fall out of wholeness?

How did we start to fall out of wholeness? How did we come to be able to do things which are 'out of harmony' with our places, our lives, our families, the planet? Has this something to do with gaining freedom or independence from God, and from our culture? The gradual separation began in ancient Greece, when abstract thinking started to become a possibility for humanity. The famous sentence attributed to Aristotle, 'All flies have six legs,' is a herald of this new achievement. This objectivity came fully to expression in the sixteenth century when the first maps were created—of the heavens, the earth and the anatomy of the human being. Imagine coming from a culture where humanity felt totally embedded in or 'at one with' nature and could only act according to the needs of nature. Imagine this, and then slowly move in evolution to a time when it became acceptable to actually cut up human bodies to look inside. This was the beginning of a real severance from the divine. We could view the earth and the heavens as something separate from us.

In this evolutionary process God or the World Spirit gradually ceases to be an integral part of our lives and starts to be something separate from us and from nature. We built mighty stone churches at a time when we could still remember God, to contain the divine, who subsequently disappeared from our fields and hedgerows, forests and furrows. If we look at the layout of landscapes prior to the sixteenth and seventeenth centuries there was still a harmony in the buildings and field boundaries that belong unmistakably to the place. A farm steading in Devon fed by lanes flanked with high-banked hedgerows, topped with oak and hazel and footed by ferns and primroses is 'of that place' and could never be transposed to the fen country of Norfolk or even the Yorkshire Dales. The way people did things was 'how it was done there'. Similarly people selected plants which suited that place, such as apples and pears, grains, root crops, cabbages, strawberries or raspberries, and created land races of crops peculiar to particular districts. They put their stamp on the plants and landscape, formed its character just as the soil and land work formed the people—in an unconscious 'belonging togetherness'. There was still an extended family of blood, while people's names, accents and even body shapes 'belonged' to where they lived. A man's ego and his land were inseparable. This integral wholeness of place and people, animals and plants allowed regional and local identity to develop. If we had gone on living like this, would we have been able to send men to the moon or develop genetically engineered crops and nuclear power stations?

We would certainly still have intact landscapes with their unique character—but what about the people? We would not have been able to separate hand craft from daily work and invent all the machines of the industrial and agricultural revolutions. The movement of our consciousness away from the earth involved a process of 'individuation'—of becoming separate from our roots. As we get a machine to do something we had previously done ourselves with our bodies or horse power—such as spinning, weaving, ploughing and harrowing—we are freed from our dependency on nature and consequently develop further faculties. We can now leave the machines to do it themselves as with automated spinning, or we pull them behind us as with modern ploughing. We no longer walk over the earth with our feet—on the contrary, we can be insulated in the cab of a tractor with headphones transporting us elsewhere!

Nature—healer of landscapes of the soul?

It is interesting that today doctors and patients alike are realizing the importance of this contact with the earth for a person's health and well-being. There is a movement 'back to the land' of people who are 'stressed-out' by modern society. What does this mean? Is the healing one way—from the land towards us—or can it also work the other way? Can we begin to give something back to an un-whole nature by 'walking over her with our feet' as Paracelsus, the Swiss sixteenth-century physician, suggested?

As science began to really take hold of western civilization, there were people in the eighteenth and nineteenth centuries who suffered deeply the loss of our connection to nature. One of these was the German novelist and playwright Johann Wolfgang von Goethe. He realized that the inspiration for his art sprang from a divine source and that this *Primal Source of All Existence* could be accessed through the study of nature. Through his efforts a path of knowledge was developed that truly acknowledged the living quality of the world and its soul-spiritual identity. Thus was Rudolf Steiner able to say:

> Goethe assumed that a single organizing principle holds sway in the world, and that this principle can be inwardly reproduced by sufficiently active and versatile thinking. In doing so, he ascribed to human knowledge the capacity not merely to observe the world externally, but to become one with the world.

Goethe's journey leads us on a path of knowledge that by its very nature aims to harmonize our world within, with that outside. This sounds to be what many are currently seeking. Is this not a path that could bring healing both for us *and* for nature?

Farming through harmonious science . . .

If we consider the image of a farmer walking over the fields on foot rather than by quad bike or '4 × 4', what happens on this daily or weekly round? The landscape of the farm is like a second skin or cloak of a farmer, who in-dwells the place. Every corner is 'known' intimately, as when a tree blew down, a stream flooded, what the movements of cows or sheep are, how pastures and soil have changed with time. When visiting a neighbouring farm, awareness of the distinctive character of a farmer's own situation becomes stronger. The way one farmer does hedge repair, fencing or stock-keeping becomes imprinted on the place. We do and feel these things more or less unconsciously, but if we want to take seriously our responsibility for our own bit of land we need first to become more aware of the healing capacity of the farmer's feet, hand, heart and head. As Peter Roth (founder of Botton Village Camphill Community) always said, 'Farmers need to think outside their welly boots'—a bit like 'thinking outside the box'. A computer buff needs a bit more 'welly' experience and the farmer perhaps a bit more 'head' thinking. Both need the heartfelt 'seeing' that Goethe's science, practised artistically and poetically, can bring us.

Science is a journey of discovery into 'what makes things tick'—a journey of 'getting to know', of uncovering the essential nature or essence of things, in effect a journey towards 'becoming one with' something other than oneself. One of the greatest scientists of our time, Albert Einstein, said the experience of 'oneness' scientists have in the 'Eureka moment', when they finally 'get it' and discover the sense of what they are seeking, they have in common with all great artists and all great lovers! Rudolf Steiner describes this moment as 'The True Communion of Man'. In the moment of 'becoming one with something', the spiritual within oneself unites with the spiritual essence of another in such a way that 'I' and 'you' disappear and there is oneness—a wholeness within a healing or holy moment. This is the aim of Goethe's science.

. . . and our poetic past

At the same time as Goethe began to put his science of 'oneness' into practice, William Wordsworth wrote the following after a walk along the River Wye above Tintern Abbey:

> For I have learned
> To look on nature, not as in the hour
> Of thoughtless youth; but hearing oftentimes

The still sad music of humanity;
Nor harsh, nor grating though of ample power,
To chasten and subdue. And I have felt
A presence that disturbs me with the joy,
Of elevated thoughts; a sense sublime; Of something far more deeply interfused,
Whose dwelling is the light of setting suns;
And the round ocean, and the living air,
And the blue sky, and in the mind of man:
A motion and a spirit, that impels all thinking things, all objects of all thought,
And rolls through all things.
Therefore I am still a lover of the meadows and the woods,
And mountains; and of all that we behold.
From this green earth; of all the mighty world
Of eye and ear, both what they half create,
And what perceive; well pleased to recognize
In nature and the language of the sense,
The anchor of my purest thoughts, the nurse,
The guide, the guardian of my heart and soul
Of all my moral being.

What is this 'motion and spirit that rolls through all things . . . and is the guide, the guardian of my heart and soul of all my moral being'?

In times gone by we took our moral guide from our parents, the church, school or guild. Even earlier the Druid priests steered their people to working in harmony with the world and the cosmos as a whole, through tuning into the stars and the *Being of the Earth Herself.*

There is an ancient Celtic invocation, which the people of the western edge of Britain used when practising the journey of unifying themselves and their place:

Christ! King of the Elements, Hear me!
Earth, Bear me
Water, Quicken me
Air, Lift me
Fire, Cleanse me.

Christ! King of the Elements, Hear me!
I will cleanse my desire through love of Thee
I will lift my heart through the air to Thee
I will offer my life renewed to Thee
I will bear the burden of earth with Thee.

Christ! King of the Elements
Fire, Water, Air and Earth
Weave within my heart this day
A cradle for Thy birth.

It came out of a recognition that the divine, which works in nature and in ourselves, has to be invoked on a regular basis in order for a process of renewal to take place.[3] It demonstrates a devotion to the *Primal Source* of all existence in every daily deed. I believe that in practice Goethe's science is very near to a modern conscious revival of the reverence found in the Celtic Christianity of our ancestors.

From Celtic Christianity to the renaissance and to spiritual science

It is interesting that soon after the dawn of reductionist science we find a rebellion in the artistic community all over Europe, of people seeking once again the divine in nature. Wordsworth was one example in Britain, while Goethe's contemporaries Phillip Otto Runge and Caspar David Friedrich were on the same quest in Germany. Meanwhile, inspired by both Goethe and Runge, Turner sought to express the essence of light and darkness in his painting. One could see Goethe as setting out on much more of a journey of discovery than the other artists—so much so that Rudolf Steiner, 100 years later and after 14 years of editing Goethe's scientific works, realized the value of the gift Goethe had brought to the world.

Out of this, Steiner developed his own journey to the spiritual depths of things as a conscious path of knowledge which he called spiritual science. This path of cognition is open to everyone who seeks a soul-spiritual dimension to their lives. In getting to know the world we get to know ourselves, have the potential to heal our inner landscapes and thereby become responsible participants in the creation of outer landscapes that are healing and beneficial to the other kingdoms of nature. The journey into the outer landscape becomes a tool for transformation of the inner self. But like the Celtic invocation, this journey passes through the four elements, through which we start to see the world with new eyes. Once this journey is embarked upon, we are able to act differently with the help of the 'Guiding Spirit of the World' recognized in the Celtic invocation as the Christ Being or Sun Spirit, with whom we are slowly able to align ourselves.

Practising 'Goetheanism' on the farm

The foregoing discussion provides a foundation for understanding Goethe's science, its relationship to the human being, its connection with biodynamics and its role at the present time in a mutually healing process between nature and humanity. In the remainder of this chapter we will look at how Goethe's science can be applied in practice.

If we now translate the journey, referred to above, into the modern practice of the farmer, and walk the same path day in, day out, week after week, year after year, the earth starts to speak. We become aware of the quality of its Being. Likewise if we watch what the animals do to the ground, touch the soil and handle the earth, we start to know the physical body of the place ('Earth, bear me'). A notebook will help to record this after asking the opinion of others—or better still, drawing and describing this in a group. *Exact sense perception* in Goethe's words[4] is like trying to describe someone you love to another person. Everyone sees the physical world from a slightly different point of view but together we build a very clear picture by viewing something from all directions of space. This takes time, days, weeks or longer of exploration, drawing and describing.

The next step, once the physical is clearly described, is to uncover something of the biography of your place. This can be done in many ways—by imagining it unfolding through the seasons of the year if you know it well, looking at old photographs or maps, talking to previous owners or local people, reading geology books and so on. The important thing is that, through all the facts, we are able to synthesize an inner moving picture of the life journey of the place that has consistency from all points of view. This 'watery seeing' becomes very lively and imaginative ('Water, quicken me'), quite different from the 'earthly' experience of facts described above. Goethe called this *exact sensorial imagination*, the key being not to fantasize beyond that which is sense-perceptible. If you have done enough basic research and are on the right track of the time process of a place, it seems to develop before the mind's eye. And if several people, especially neighbours or previous owners, describe the same journey from another point of view, this is tremendously affirming and enlightening for one's own experience.

By living into these 'time' processes, certain patterns, revelations or realizations start to appear—the character of the place of our farm or one of its 'rooms' begins to emerge and speak to us. It is as if we lift out of and above the farm and can see the whole for the first time ('Air, lift me'). We get a glimpse of a mood or atmosphere that belongs to this specific place and to no other. This is what makes it unique. The *genius loci* starts to shine. Spinoza, the Italian-Dutch philosopher whose approach Goethe found closest to his own, described this type of knowing as when 'certain attributes of God make themselves known within the things'. Goethe called it '*The power to judge in beholding'*. These are moments of grace when one 'sees' as if for the first time—'I realize' or 'recognize' something already known from another time or place.

Taking this further affirms it and allows a *becoming one with* or true *knowing* to become the basis of one's deeds. That 'spirit, which rolls through all things' (Wordsworth) has then truly become 'the guide, the guardian of my heart and soul, of all my moral being'. My own wishes and desires disappear. I feel cleansed, fired up. The morality indwelling the earth has touched me. Such moments are experienced as deeply humbling and very near to a religious revelation. What is truly wonderful is that anyone can have such an experience if they are open to it. When they happen within a group of people it is like a communion with the Being of the living Earth. This forms the basis of *consensus design*—creative action based on a moral revelation after a scientific journey. Here, individual egos disappear and the resultant acts are energized by a sense of service to the highest Ego common to all of humanity. This, in the Christian tradition, is expressed as 'Not I, but Christ in me'.[5]

'Not I, but Christ in me'!

At this point we have come to the central part of the inner and outer journey into landscape—into ourselves. The outside of the landscape becomes the inside of me and I can start to 'live into it' as if it were an intimate lover. If we manage this, we may wake up in the morning after being full of questions the night before, 'knowing' exactly what to do to solve a particular problem. We start to be able to 'think' with our hands and feet. Head and heart are participating in the 'welly boot walking' and, what is more, we know how we reached this or that decision. Our farming has become near to a religious or holy practice.

In this twenty-first century we are now striving to find meaning, to be able once again to call things by their true 'names'. We struggle to find words to express ourselves when graced with a glimpse of the spiritual essence of another being. We strive to purge or purify ourselves in order to be able to 'hear' the other speak its essence within us. This act of working on oneself to be able to 'tune into another being' is healing for ourselves and the world we encounter. Our inner landscape is undergoing transformation and this will find its echo in the outer landscape of our farm or garden. Perhaps this is the gift we have to offer the world of nature when we learn to work in a Goethean way.

Building places which can shine

I believe this method is the foundation of a new morality—a morality based on the integrity of the earth. Just as when we are really given the space to be ourselves by another person we feel more than we are, and can grow, so does the landscape or all of nature. This is perhaps what is meant by 'walking with Christ'—allowing space for the divine in nature to once again actively participate in our lives. The middle four lines of the Celtic invocation reveal this. Is this not a healing process both for us and our landscapes?

In this way a group of people under the auspices of the Life Science Trust has 'journeyed' together in order to study and transform 60 acres of land at the foot of the Lammermuir Hills in south-east Scotland—the Pishwanton Wood Project[6] (Fig. 9.1). Here, Goethean science has been used in the service of landscape, architecture and the harmonization of place, within the context of the wider East Lothian landscape. The process, as explained above, has involved penetrating the layers of landscape via the four elements.

Earth We begin with rock, the earth, which we meet outwardly with our senses.
We meet the world in its outer manifestation with the outer part of ourselves—our skin and other sense organs.

Water Hovering around and weaving through this factual encounter is the changing, transforming nature of the day, the year, of everything which grows and develops. Then we move on to connect ourselves with the fluidity of a place in its time process—that which weaves in and between all the component parts.

Air Experiencing the biography of a place through reliving its time process inwardly is coloured by our own likes and dislikes, our soul moods mixed with the moods emanating from whatever we meet. If we have built a solid foundation as we move along this path, it becomes obvious which is me and which is my place speaking. Hidden in all this we find revealed—like an airborne flash—the true nature of the place, the *genius loci*.

Fire Hidden deep under all these layers lies the 'essential being' of a plant, of a place or person—that which we know and can name. Going on this journey thus far has the effect of uniting us intimately with a place, such that we have intuitively 'become one with it' while at the same time still having our feet firmly on the ground. All of these elements woven together are what we recognize with the different levels of our own being as 'landscape'. We have then touched something so profound that our lives can be changed and our creativity transformed—for whatever we subsequently do there.

Fig. 9.1 Pishwanton's animals and Lammerlaw

If we return to our example of the farmer walking the land, once embarked on this journey of self-development in relation to the outer landscape we find the way we do things will change. Decisions seem to emerge from the depths of our being, as if the landscape is speaking within us. We find we 'know' what to do and that other people working beside us will have the same idea or picture from their point of view. This, as we earlier indicated, is the basis of consensus design[7]—not an agreement with one another as an attempt to converge or compromise, but rather an enhanced 'seeing' that comes out of the spiritual essence of things. We know we have reached this point when there is a common picture—led by the landscape itself.

If the biography of a place has been fully studied and inwardly imagined and a picture of the future intentions has been clearly visualized, then there can often be a seamless flow from past to future in the working out of an intention. Thus the craft workshop at Pishwanton (Fig. 9.2), designed and built between 2000 and 2002, looks as if it has grown out of

Fig. 9.2 The Pishwanton craft workshop

the ground, while its undulating roof echoes the hills behind. Within the landscape as a whole, back in 1991 we found something of the gesture of the human organism. This is described in Christopher Day's *Consensus Design* along with details of the 'architectural journey' including choice of field boundaries, woodland planting, gates and footpaths.

In the 'central bowl'—the then proposed garden area—we discovered again the gesture of the human organism. At one end was a quiet meditative area asking for seats and inner contemplation under the pines. The other end had much activity, springs, heavy clay soil and plants healing for the lower human being. In the middle was a rippling, breathing area with interpenetrating path flow. We felt this was asking to be enhanced, so now the planting scheme incorporates healing herbs for all the different parts of the body. But in outline, Pishwanton's herb garden (Figs 9.3 and 9.4) was 'seen' in its current place 17 years before it

Fig. 9.3 Pishwanton herb garden

Fig. 9.4 Close-up of herb garden bed design

Fig. 9.5 The Goethean Study Centre

came into being while the undulations of the surrounding hills and craft workshop roof have found their way into its rippling raised beds!

Finally, the Goethean Study Centre and herb processing facility (Fig. 9.5) has a characteristic 'organic' structure, which reflects the nature of the place in the context of the whole. It is like a human head resting on an oak ring-beam with lots of space inside for thinking. Compare this with the craft workshop—a space for limb work built of strong stone on solid foundations—and the garden, a heart and breathing space.

As landscape development is an on-going process at Pishwanton, any who would like to visit or participate are most welcome.[8]

10. Food Quality, Nutrition and Health

Our main concern in this chapter is to show that organic and biodynamic food has an inherent quality which supports health, and why that should be the case. Food is obviously fundamental to our health, but how easily can society's health problems be linked to food? To begin with, it will be helpful to explore the relationship between diet and health and to develop a picture of what is, or is not, a genuine indication of food quality. After considering evidence for the quality of products and diets based on organic and biodynamic food, we will probe unfamiliar nutritional concepts arising from the insights of Rudolf Steiner. It may then be possible to offer a perspective on some of society's current health problems.

Diet and health: a problematic linkage

What is health? Sir Albert Howard considered it to be the 'birthright of all living things'. It is sometimes answered in the negative, as the absence of disease or injury! But as an actual human experience, it corresponds to a feeling of physical and psychological well-being. In the dictionary, the word is seen to derive from 'wholeness'.

We can speak of the health of the individual, and the health of society. Periodically voices are heard to proclaim that society's health is better now than 20–30 years ago, yet there is a higher incidence of stress-related and eating disorders than ever before and obesity has recently joined cancers, cardiovascular disease and diabetes as an epidemic of modern western life (Anglo-Saxon at least). A cynic might say we are being kept alive by healthcare systems and advances in medical treatment. Furthermore, if one embraces the definition of health already offered, behaviour should be included with society's health, for that is clearly an extension of the psychological outlook of human beings and that alone gives cause for deep concern. The point of mentioning this spectrum of issues, physical and psychological or spiritual, is that we are definitely at fault if we narrow the discussion of health to physiological questions alone.

Health is strongly affected by the aggregate of our food intake—our diet. At one extreme, the malnourished child will be prone to illness, while the affluent and over-indulgent place themselves at other sorts of risk. This is no trivial issue, for we have become victims of food addiction, as those trying to influence the diet of youngsters must acknowledge.

Yet health issues are complex. One can think of diseases like cholera and hepatitis which have a clearly identified pathogenic origin, but the majority of diseases have multiple aetiology—cancer being a notorious example—which makes pinning down a primary origin extremely difficult or perhaps meaningless. Many risk factors are routinely identified as contributing to ill health including genetic predisposition, lifestyle, and workplace or environmental hazards. Diet has often been an afterthought. The socio-economic factor alone reveals that disposable income affects access to food choices. As regards diet in the UK, it is worth reflecting that in relation to money spent on cancer treatment per head of population we experience among the lowest rates of recovery in Europe.

Because of each person's different background and mobility in the course of life, the effect on health of a balanced diet or of quality food has remained largely obscured. In addition, we are supplied with a stream of differing and often conflicting advice for avoidance or alleviation of health problems. While this has led to many pursuing what they believe to be a positive approach to health, it has simultaneously confused and deterred many from taking any steps at all away from 'eating as usual'. This has all been much to the benefit of the food industry which has been free to add value to its products, mostly with ignorance of the possible consequences for the health of consumers.[1]

Today, both food and cures are in the hands of powerful bodies. While consumer relations ensure a minimum level of response to risk factors, the commercial world is nervous about the public having a deeper knowledge of diet and health. It is easier for big corporations to deal with humanity as if it were a herd of cattle. And the last thing wanted by drug companies—and perhaps also by orthodox medicine—is to be faced with inexpensive alternatives.

Food quality as understood in the marketplace

For the customer, the primary signal of quality is visual, so from the appearance of fruit and vegetables we detect freshness and other aesthetic characteristics such as colour and absence of blemishes. But as with human beings, appearances can be deceptive! Fruit can be artificially glazed to enhance its appearance while food in general gives no outward sign of substances likely to cause harm. We can touch and smell loose items, which provides us with additional signals of freshness or ripeness, but for packaged goods we must rely on appearance alone, the 'sealed-in' freshness assured by the wrapping, and of course, the 'best before' date. EU trading rules, together with supermarket preferences for packaging and presentation, have led to standardization of size and shape in whole ranges of fruits and vegetables. This has created additional pressures for producers, has increased consumption of packaging and has led to a narrowing of what customers are prepared to accept.

Of course, it is usually not until we get home that we can taste the food to determine whether an apple has flavour, how tough its skin is, whether it has a crisp or mealy texture or whether the contents of that avocado are discoloured and inedible. Indeed, this raises the issue of the long storage to which much of our fruit has been subject prior to its display in shops, together with the artificial ripening which may have been carried out. For this reason, aside from seasonal produce, 'fresh fruit' has become just one of many deceptions.

Milk, too, is certainly bought 'fresh' and for historically sound reasons is pasteurized. But much milk is now homogenized simply to extend shelf-life. It is available whole or with varying amounts of natural fats skimmed off. Milk is therefore a largely denatured, processed food and can, in the UK at least, only be obtained raw (green top) from farms with a private milk round or direct sales.

Processes to improve appearance can have real benefits for the consumer—the power-washing of potatoes and carrots for example. While targeting the shopper attracted by clean appearance or simpler preparation, this immediately reveals the produce to be free from scab, blight or carrot fly, all of which might otherwise be obscured by soil! On the other hand, this kind of treatment is likely to reduce the shelf-life of produce. Meanwhile, if we consider the treatment of most meat and its processed variants, it appears that British customers are offered the artificial red colour they are said to prefer. So the retailing of food has generated bizarre notions of quality by way of customer appeal.

As the shopper so evidently operates on the basis of quality by proxy, it might be of equal or greater interest to consider how food is *produced* rather than how it is *presented*. Here, the choice is between 'most produce' and 'organic produce', with labels such as 'free-range' or 'natural' representing a kind of half-way house. In the face of price comparisons, the decision to buy organic is based on a combination of factors including concerns over pesticide or antibiotic residues, genetic modification, the environment and the treatment of animals. Increasing numbers of people are also becoming aware of the health benefits of organic food,[2] especially with the rise of food intolerance. This means that the majority will purchase organic food because of what it does *not* contain or involve, rather than through any awareness of its innate characteristics.

Organic and biodynamic food quality

Many are inclined to feel that 'organic' is merely a marketing ploy and that there is no real evidence of its superior quality. In this respect, a common tendency is to adopt the views or attitudes of others without bothering to make serious enquiries. Anyone wishing to malign organics can therefore be sure that their remarks will be perpetuated, particularly as the real value of organic and biodynamic food, as we shall see, cannot be measured by conventional tests alone.

It is therefore useful to start with a reminder that food safety has become a major concern in recent years—with meat products as well as vegetables and fruit. Organic produce, through its regulatory process, offers at the very least an assurance concerning residues of pesticides, antibiotics and GM products. People suffering from allergies gain relief through confining themselves to organic produce while for many years it has been the cornerstone of alternative cancer therapy.[3] For this reason, that occasional organic lettuce which has a living occupant should be greeted with relief rather than disdain!

Of course, as eating is a human experience, many of the reports about organic and biodynamic (Demeter) produce are subjective and are therefore inclined to be dismissed by those looking for 'hard' evidence. Many report that such food has better flavour, aroma and cooking characteristics.[4] Top chefs are now among those prepared to put their heads above the parapet—even to giving top marks to biodynamic produce. This writer is not alone in claiming that organic produce—bread for example—gives more sustained satisfaction from small quantities, an interesting question in view of complaints about the cost of organic food. He has also encountered reports of organic beef cattle requiring less than half the amount of supplementary grain in wintertime as compared to non-organic herds.

It is perhaps ironical that biodynamics has become widely known for the quality of wine it produces—and here there really are experts able to judge the best from the average! Following the thread of Chapter 3, biodynamics is uniquely able to foster the local characteristics (*terroir*) of produce. This helps explain why the wine marketed as 'organic' in France is increasingly biodynamic. It has also been declared that the taste of wine is better on 'fruit days' of the biodynamic calendar, though whether this is wholly a function of the wine is not clear!

If anything, there is wider acclaim for the flavour of organic meat than with fruit and vegetables. Independent tasting panels report that two critical factors determining flavour in meat, aside from animal breed, are the quality of feed and levels of stress.[5] While strict feed quality rules apply on organic and biodynamic farms, animals experience little stress compared with intensive systems. This results in minimal veterinary costs.

Nutritional characteristics and keeping quality

As we reported in Chapter 1, organic produce has smaller cells with thicker cell walls. It therefore mostly contains less water per unit volume than conventionally grown produce. The flavour and texture of organic fruit and vegetables will to some extent be traced to these characteristics and to the content of minerals and other plant chemicals. A number of analytical studies show organic and biodynamic produce to contain higher levels of particular nutrients and vitamins. Vitamin C and iron are notable, while calcium, magnesium, potassium and phosphorus tend to exceed non-organic comparisons. The single set of data below shows some of

Nutritional differences between organic and non-organic produce. Weights per 100 g. All comparisons significant at 95%. (After Pither and Hall, MAFF, 1990)

Apples	**Organic**	**Non-organic**
Total sugars	8.8 g	9.5 g
Vitamin C	21.6 mg	19.3 mg
Tomatoes		
Vitamin C	21.8 mg	18.0 mg
Vitamin A	4.7 mg	3.5 mg
Carrots		
Glucose	0.9 g	1.3 g
Potassium	269 mg	217 mg
Potatoes		
Total sugars	0.7 g	0.8 g
Vitamin C	13.5 mg	17.8 mg
Potassium	329 mg	370 mg
Zinc	310 μg	260 μg
Potatoes after dehydration		
Sucrose	1.0 g	2.4 g
Fructose	1.2 g	0.7 g
Glucose	2.0 g	1.2 g
Iron	5.7 mg	4.7 mg
Calcium	64.0 mg	56.4 mg
Zinc	1810 μg	1350 μg

these characteristics. On the other hand, nitrates are significantly down as compared to non-organic produce. Organic plant products often display a higher natural sugar content and higher levels of accessory organic substances which are strongly associated with flavour. A substantial proportion of the latter have antioxidant properties. A study at Newcastle University has recently shown antioxidant levels to be 90% higher in organic than non-organic milk and for organic vegetables they were up to 40% higher. Significantly, during the period throughout which chemical fertilizers and pesticides made their greatest impact on the food market there was a 75% reduction in the mineral content of fruit and vegetables.[6] This, as we shall see later, appears to have incurred its own toll on health, and certainly helps explain why so many today are taking mineral supplements.

Better known to growers and retailers than to consumers is the fact that organic produce has superior keeping quality. This might not seem such an important property except for what it reveals about the contents! Biodynamic fruit from Australia can be shipped in a riper state without fear of damage, organic onions in northern Ghana remain saleable for four months while conventional ones two at most. Pettersson's study displayed below showed that potatoes grown conventionally achieved 4 tonnes per hectare more than a paired biodynamic crop, yet after several months

Yields and quality of potatoes under biodynamic and non-organic management (after Pettersson, 1977)

Measurements	Biodynamic	Non-organic
Yields, October (t/ha)	34.2	38.2
Weight, April (t/ha)	30.0	26.6
Losses by grading and storage (%)	12.5	30.2
Crude protein (% of dry matter)	7.7	10.4
True protein (% of crude protein)	65.8	61.0
Vitamin C (mg/100g fresh matter)	18.1	15.5
Darkening of extract (E.10^3, 48h 8°C)	354	462
Decomposition of extract (EC)	22.0	30.9
Crystallization defects	4.2	5.2
Taste points (best = 4)		
December	3.1	3.0
April	2.7	2.3
Cooking defects		
December	1.8	4.1
April	2.1	9.2

storage the situation was reversed through higher levels of wastage incurred by the conventional crop.[7]

What is it that causes two equally fresh products to behave so differently? In a world which is facing real food shortages, any post-harvest losses will need to be examined in a way that has never appeared necessary before. We might suspect that the nutritional character of organic products, particularly their content of cyanogenic glucosides and other natural defence substances, plays a part. Such substances might inhibit the production of decomposition enzymes and thus the proliferation of bacteria, but even so standard nutritional comparisons between 'organic' and 'non-organic' are often not far apart. Insight into this question is provided by studies showing improved keeping quality according to use of the biodynamic preparations. For example, Samaras' results, shown below, imply that biodynamic practices strengthen that inherent aspect of produce quality which we have referred to elsewhere as the 'life force'.

Organic diets and health

A variety of experiments with animals, mostly rats, rabbits and hens,[8] reveal more about the quality of an organic diet. When offered a choice between organic and non-organic feed, in virtually all tests the animals chose the organic option. It has been the same when farm animals are

The effect of biodynamic preparations on storage quality of carrots (after Samaras, 1978)

CM = manure composted with preparations 502–507, 500 = horn manure, 501 = horn silica

Measurements	**CM alone**	**CM + 3 × 500**	**CM + 3 × 500 + 4 × 501**
CO_2 mg/kg fresh weight after 94 h	2831	2755	2565
Catalase activity/g dry matter	397	375	346
Peroxidase activity/g dry matter	15992	11051	7591
Saccharase activity/g dry matter	90	98	112
Amylase activity/g dry matter	72	75	81
Bacteria count, M/g fresh weight	1.67	0.45	0.43
Fungi count, M/g fresh weight	2.7	2.7	2.6
Loss in weight, % dry matter	56.1	46.6	29.2
Spoilage, % after 164 days	28.2	23.0	20.4

offered rainwater or water from the public supply—they will choose the more natural supply. In these preference tests animals clearly select the food that instinct tells them is more beneficial to their health.

Further experiments have involved feeding separate groups of animals organic or non-organic diets and observing their health and behaviour over several generations. Among those which can be properly validated, a pattern has emerged of sustained vitality in subsequent generations among the organic group while health and reproductive capacity—expressed as sperm activity or ovulation rhythms—markedly deteriorates among the others.[9] Exposure to pesticides could explain part of this picture[10] but the 'generational effect' suggests that non-organic progeny may be *inheriting* reduced metabolic capacities.[11] This idea is supported by research indicating that health may be linked to conditions in the grandparental generation.[12]

As yet, we are at an early stage in the assessment of food on human behaviour but certain indications should nonetheless be reported. Recent British experience suggests that discouraging certain drinks with artificial colour reduces attention-deficit/hyperactivity disorder (ADHD) in schools while a study over four years in New York showed that a group of children 6–7 points below average IQ ended up well over average as a result of keeping off 'junk food'. Again, a study in Dornach, Switzerland, showed that a group of students sustained their ability in memory and mathematical tests after eating biodynamic food while they significantly failed to do so after eating normal, non-organic food. A study by Essex University also reveals that Jamie Oliver's 'Feed me Better' schools food campaign is having an impact on attendance, concentration and examination results.[13] The author has also learnt of a report from central Africa where a school which was set up to feed students with organic food from the school garden had such good exam results that government inspectors insisted the children repeat the exam elsewhere. They did so, and with the same outcome! While this may largely be related to a better diet we can reflect on how important quality food is for the young in particular. However, Huber and colleagues' recent study of nuns in a German convent provides one of the best examples to date of the benefits of organic food on human health.[14]

In the comparison between organic and non-organic, are we merely dealing with the depressive effects of pesticides and other foreign materials—with nutritional differences? Or is there some additional inherent vitality associated with organic food and organically fed animals? The foregoing certainly leads us to conclude there is. In the animals' food preference tests it should be made clear that they do not go back and forth sampling the alternatives before commencing to eat, but react to the subtle energies which all natural products contain. As this is a difficult idea for many people, it is important we look for independent evidence to corroborate this extra vitality or subtle energy.

Life forces and their survival in food

As we said in Chapter 4, it is not possible to analyse 'life', only to observe or monitor its processes—in other words to see its effects. Since the earlier years of the twentieth century, research has been carried out into picture-forming methods that are understood to express the formative gestures of living processes.[15] The first of these is the capillary chromatographic method of Eugen and Lili Kolisko.[16] Here, a biologically active solution is allowed to permeate chromatographic paper previously impregnated with the photo-sensitive material silver nitrate. Instead of forming simple bands as in normal paper chromatography, the outcome is a pattern of waves, folds or flamelike structures (Fig. 10.1). To the original cylindrical pictures circular chromatograms were added,[17] in this instance comparing carrots grown by different methods (Fig. 10.2).

There are two aspects of this method which we need to recognize. Firstly, the patterns vary in intensity and reveal a pseudo-three-dimensional character. Such could arise only if a formative energy was radiating from the material while the liquid carrying this energy moved up or across the paper by capillarity. Secondly, the patterns are reminiscent of plant forms and as such accord with Steiner's remarks that the forces that animate life are those of the plant world.

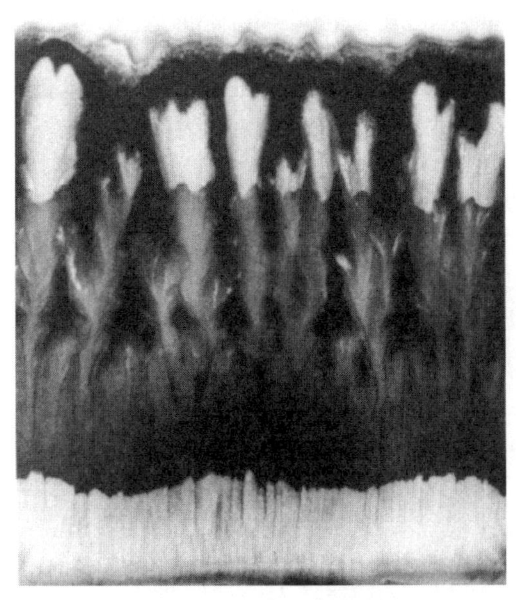

Fig. 10.1 Cow's urine—a 'capillary dynamolysis' picture (from Kolisko, 1978)

A further technique to develop along these lines was sensitive crystallization or biocrystallization.[18] Here, the

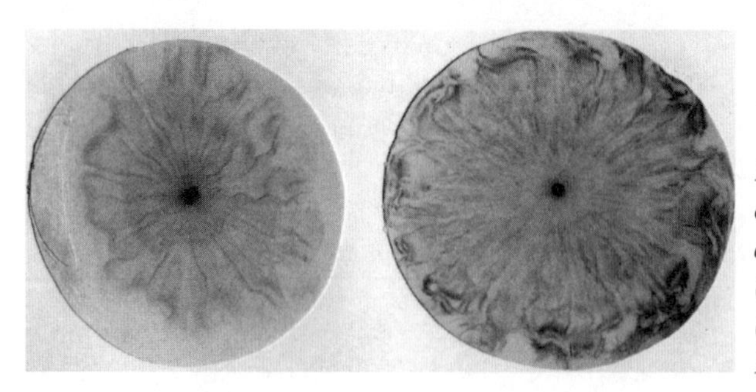

Fig. 10.2 Freshly extracted juice from carrots. Conventional (left) and biodynamic (right). (A. and M. Kleinjans)

characteristic of mineral salts to form crystals is combined with the formative influences of a biologically active material. Thus when copper chloride is allowed to crystallize under controlled conditions we obtain the pattern shown on the left in Fig. 10.3. If an organic extract is introduced into the same solution (Fig. 10.3 right) crystallization proceeds according to different principles, with plantlike forms displayed. Because this technique has attained a relatively high level of reproducibility it has been used to differentiate fresh from aged samples, healthy from unhealthy organisms, traditional varieties from modern, and organic from non-organic samples.[19]

A point rarely mentioned when reviewing these techniques is that they are affected by stronger cosmic events such as eclipses and moon at perigee[20] (Fig. 6.9). While this may be seized upon as evidence of non-reproducibility and therefore of unreliability, it tends to confirm rather than dismiss the notion that life energies actually do radiate from biological materials. As life energies are initially drawn from the cosmos, we might expect these primary influences to affect the way life forces are expressed under experimental conditions.

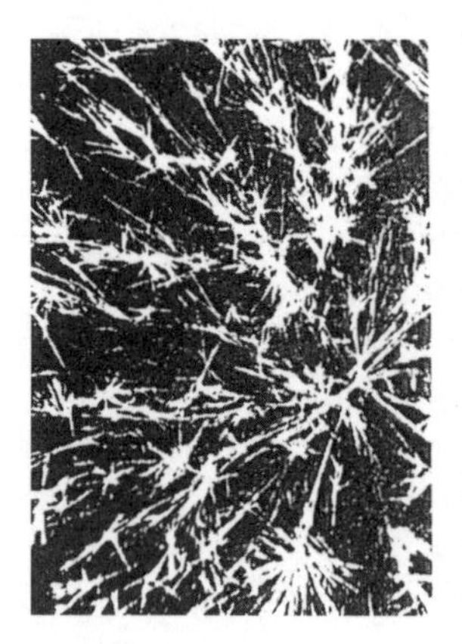

Fig. 10.3 Sensitive crystallization. Pure copper chloride crystals (left) and copper chloride crystallized with extract of wheat germs. (From Koepf, 1993)

Whilst 'vital forces' will remain a controversial issue among a wider public there is nevertheless agreement that certain factors lead to a deterioration of food quality. Not surprisingly, vitality as assessed by capillary and crystallization methods will generally decline with storage time, by overcooking and by different forms of food processing. In addition to the example in Fig. 10.2 let us consider the comparisons demonstrated by Kolisko (Fig. 10.4) and a more recent crystallization study (Fig. 10.5). In all such examples, fresh material has more pronounced formative characteristics which we should take as evidence of greater vital force. Milk homogenization also appears to impair its quality[21] (Fig. 10.6).

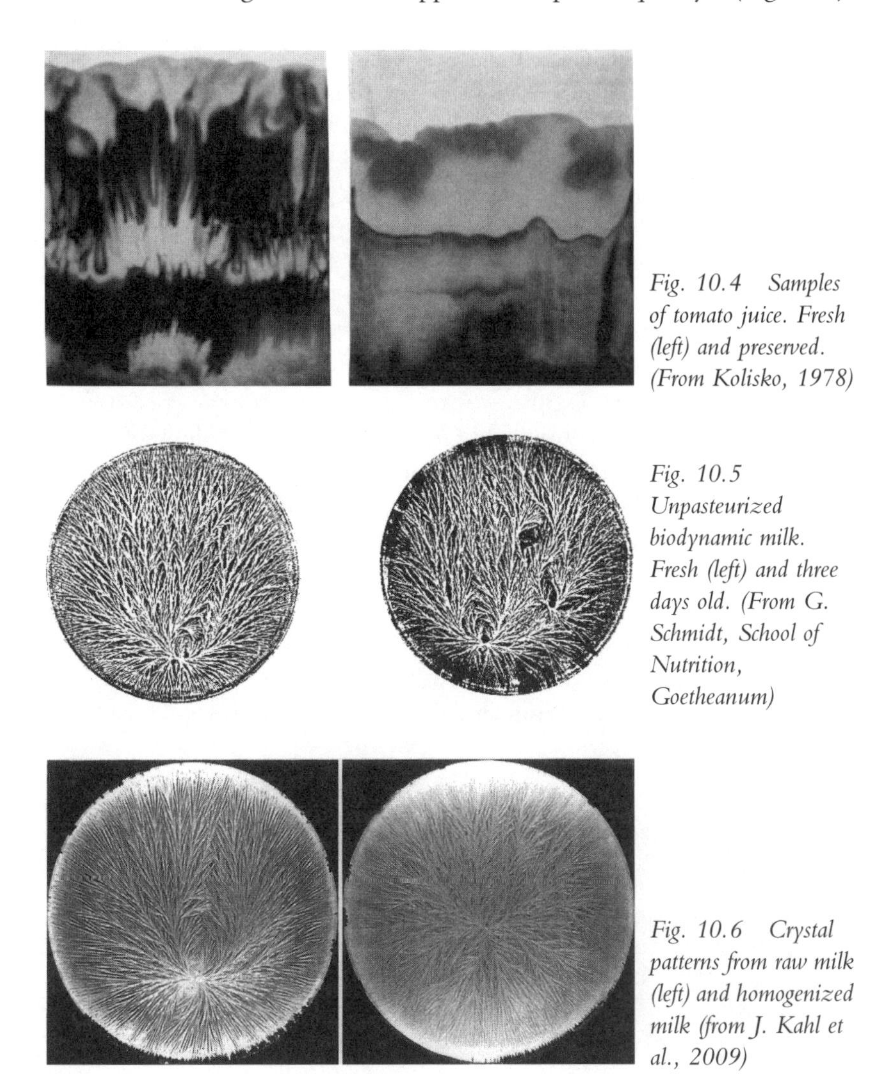

Fig. 10.4 Samples of tomato juice. Fresh (left) and preserved. (From Kolisko, 1978)

Fig. 10.5 Unpasteurized biodynamic milk. Fresh (left) and three days old. (From G. Schmidt, School of Nutrition, Goetheanum)

Fig. 10.6 Crystal patterns from raw milk (left) and homogenized milk (from J. Kahl et al., 2009)

It is widely accepted that overcooking spoils the taste of food and that vitamins are lost in the process. It is also general knowledge that the flower and leaf part of plants requires relatively little cooking while roots need longer. Hauschka conducted chromatographic tests on food samples to demonstrate that only the shortest cooking times should be employed for vegetables like cauliflower, otherwise the vital forces would be lost.[22] Similarly, types of preserving which employ the minimum of heat are the most successful for maintaining a level of life force. Cooking, as a heat process, is equivalent to ripening, as it raises the food to a state where it is less effort to digest—an issue touched on in the following section. In this way, foods cooked appropriately provide good nourishment. However, according to Alan Hall,[23] when food is microwaved 'all its vibration patterns go'—his code for vital forces! Again, Hauschka's germination tests showed that such is the sensitivity of water that even the material of cooking utensils has an influence on the food being prepared.[24]

The industrial processing of food involves varying degrees of denaturing and reconstitution so is generally bad news from the point of view of vital forces. Even when considering polished rice, we have lost a protein-rich outer layer and are left mainly with starch, this having implications for the strength of formative forces (Fig. 10.7).[25] The creation of cereal flakes would appear relatively benign, while refinement of sugar and flours represents a movement away from the spectrum of substances present in the original 'whole food'. Even less desirable is the production of margarines and items containing numerous additives. These can be regarded as 'dead food' whose content is set to deceive rather than nourish the human being, despite being 'fortified by vitamins' and free from substances currently out of favour with doctors and nutritionists.

An indication of just how deficient are processed foods was given long ago by McCarrison's experiments with rats. Those offered a diet containing live or natural foods remained healthy while others fed a western-

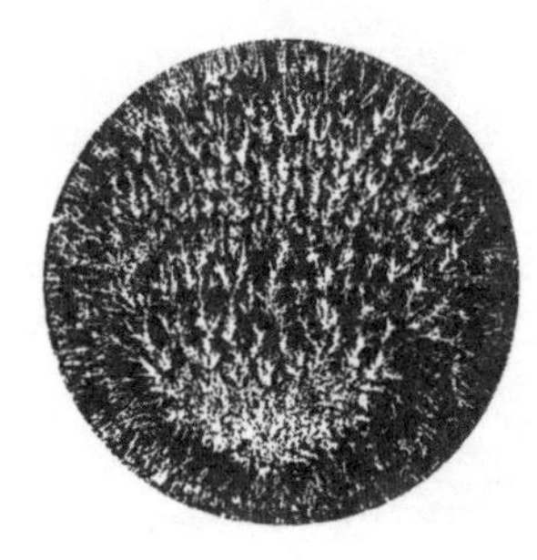

Fig. 10.7 Samples of rice. Whole grain (left) and polished. (From U. Balzer-Graf)

style diet not only developed poor health but their behaviour changed. They appeared to lose their natural instincts and started attacking each other.[26] Indeed, analyses carried out in the 1960s showed that the cardboard packaging of certain puffed cereals was scarcely less nutritious than the contents! Further insight is drawn from Bunge's experiment which fed mice with milk or an alternative that was synthesized from the separate substances milk contains. Those fed the latter all died after several days.[27]

In the 1970s it was discovered that living cells and organisms have the characteristic of spontaneously emitting low levels of light. Fritz Popp and subsequent researchers have thus made reference to 'biophotons'. Such work is aimed at establishing the 'coherence' of organisms, namely that a living organism is more than the sum total of its cellular make-up. From studies of photon emissions it has been possible to distinguish organic from non-organic tomatoes, free-range from battery eggs, and to predict the germination rate of seeds from their afterglow.[28]

In continuing our journey with food and health, we should now try to build a picture of why quality in terms of 'vital forces' or 'coherence' should be of such importance for the human digestive and nutritional process.

Alternative concepts in nutrition

A fundamental point of divergence between the picture of nutrition offered by Rudolf Steiner and the orthodox one is the principle that we need food for the life-giving *forces* it contains as well as for its physical *substances*. This is why the issue of whether food contains these forces is of such importance and why special account has been taken of their existence in the foregoing section. Organic and biodynamic foods are likely to contain these forces in greater measure than food 'conventionally' produced for reasons explained in earlier chapters. But we can so overcook food of whatever origin, or cook it in an inappropriate way, that vital energies or forces are driven out. Meanwhile, many processed foods have little prospect of offering us 'forces' in the first place!

Today it is commonly assumed that the less effort needed to digest foods the better, albeit we do it unconsciously. But while effort expended *outwardly* in physical activity tires the body, effort expended *inwardly* releases vital energies. Steiner explained that the benefit we gain from food's inner forces is proportional to the effort expended in digestion, and food that contains vital forces involves greater digestive effort on our part.

Before considering the process of digestion in more detail, a brief point should be made about meat versus vegetarian diets. According to Steiner, as compared with vegetable protein, animal protein requires less effort to digest. So, for the above reason, we gain less benefit from it. And since it contains a residue of the animal's astrality, it can 'burden' the body in a way that vegetable protein does not.[29] This may be a contributory reason for why after a heavy meal we can feel drowsy. Vegetarian diets can therefore be recommended for giving us greater vitality without 'side effects'. It will also be better understood why raw foods—usually vegetables—can be therapeutic through releasing the highest levels of vital forces, together with intact vitamins and a complete inventory of salts.

Earthly and cosmic nutritional pathways

The orthodox view of nutrition considers that substances taken in as food contribute energy and raw materials to maintain bodily processes. Digested materials are transferred through the stomach or intestinal wall and thence into the lymphatic and blood stream. For Steiner, this was essentially a picture of maintaining a machine by periodic refuelling. Within a radical and over-arching framework, he distinguished between *earthly* and *cosmic* nutrition in animals and human beings. This might appear absurd, yet no one could seriously doubt the working together of cosmic and earthly aspects in *plant* nutrition. So in this conception we have material and non-material influences interacting, and what is eaten has both a direct and indirect relationship to bodily processes.[30]

The food and liquid we consume—earthly nutrition—has a complex but broadly threefold function. It firstly provides, via carbohydrates and fats, fuel and warmth for the body to conduct its various tasks—principally providing energy which is immediately or more slowly available.[31] The solar origin of this energy was discussed in Chapter 4. Meanwhile, the body requires liquid intake for maintaining its vital functions and to facilitate the excretory process.

Secondly, with a vital contribution from mineral salts,[32] it nourishes and provides a physical basis for the nervous system. While built from earthly substance, the capacity of the nerve-sense system to function depends on the reception and deployment of cosmic forces for which the brain gives direction. In this way, the head directs the functioning of bodily processes as well as our higher capacities. Earthly nutrition therefore needs to build a strong nervous system so that effective influence can be directed from the head.

Thirdly, while much of the bulk of what we eat is excreted, the digestive process brings it about that beyond the key digestive organs a force or spiritual essence is released into our etheric circulation. Thus, after digestion, carbohydrates, in addition to their release of energy, provide the body with what it needs to create and maintain bodily *form*[33] while fats have a relationship to the filling of form with *substance*. We have a distinct analogue here between the roles of silica and calcium as discussed in Chapter 2. None of these processes would be possible, however, without the existence of proteins.

From the proteins in our everyday food, an essence is gathered from which our body forms its *own* protein. The protein we eat should be completely destroyed by our digestive system, this work being concluded in the intestines. An essence of this protein—let us call it a blueprint—passes through the intestinal wall[34] and into the body's etheric or life energy system. Here it encounters 'cosmic nutrition' which has been transmitted by the nerves. Thus an essential part of our nutrition comes from outside us and is drawn in through our senses. It comes in through all sense organs, the skin, the eyes and, in the case of oxygen and nitrogen, through our breathing.[35] Let us recall that the human being is an image of the cosmos and that our organs are shaped by cosmic forces (Chapter 2).

Our individual senses connect on an unconscious level with the various ethers or principles underlying matter. Just as we consciously experience sensations of light, colour and sound, we continually receive impressions from the elemental world which produce 'imaginations' in our etheric body.[36] In this way cosmic forces pass to the inner organs where, using the blueprint distributed by the vital body, new substance is condensed, allowing cell replenishment to take place. In Hauschka's words, 'our senses reach out to perceive and absorb spiritual sustenance in the form of creative forces proceeding from the divine cosmos. Like a plant, this grows down from the head out of spirit into matter, condensing as flesh and blood'.[37]

For all those struggling with the idea of non-material aspects participating in digestion, consider how important are sight and smell as precursors to digestion. It will then be appreciated how connected are the various aspects of our organism—astral, etheric and physical—in the overall process. The digestive-nutritional process is thus one of unbelievable accomplishment, with the ascent of earthly substance engaging with the descent of influences from the cosmos. Cosmic forces strike firstly into our astral organism (the sentient body), physically represented by the nervous system. They then unite with forces absorbed into the etheric body from the food we have eaten. In a pivotal position between incoming cosmic forces and our vital or etheric organism are the endocrine glands, which have the task of regulation.[38]

We have periodically noted that plants and animals draw different aspects of their life process from the inner and outer planets. Steiner emphasized that if our food does not contain a proper balance of planetary forces the formation of our body's own protein will not be conducted in a harmonious fashion. Thus we can either have a blueprint which looks blurred, and which may have some bits missing, or one which has sharper focus.[39] The implications of these alternatives must be clear. On this basis, as we explained in Chapter 4, most agricultural produce worldwide lacks a balanced profile of planetary forces, thereby limiting its value in the support of animal and human health. As has been consistently commented, achieving that balance is a major aim of biodynamic agriculture.

So let us be quite clear why the quality of our food is of such importance. Quality matters firstly because it determines the body's capacity to receive incoming cosmic influences. It matters secondly because the quality of the de-materialized element from our food affects our ability accurately to repair and replenish bodily substance.[40] Thirdly, through achieving a proper balance between these principles, the body can maintain an effective regulatory system incorporating endocrine and immune system functions. It follows that a true definition of quality food is: *that which enables a proper working of the different elements of our nutritional system—earthly and cosmic.*

Nutritional relationships between plants and the human being

Difficult as these nutritional ideas certainly are, there is a further important relationship to consider. In many of his lectures, Rudolf Steiner spoke of plants and human beings as exhibiting a broadly threefold division (Fig. 10.8). The plant root connects with the nerve-sense system—directed by the head. Stems and leaves stimulate the rhythmic system of heart and lungs, from diaphragm upwards, while flowers and fruit influence the lower metabolic and reproductive parts. Seeds encapsulate the whole plant and as such are nourishment for the entire body. The human being can therefore be considered an upside-down plant, or vice-versa.

Although these different plant parts are associated with a varying balance of nutritional components, it is the forces or subtle energies each carries which are needed to maintain the health of the body.[41] This is an important step towards building a healthier nutrition and treating a range of illnesses. Here, we can be reminded that the body's etheric or life forces are plantlike in character, as reflected in the plantlike images of capillary and crystallization pictures. It is the same with the herbal preparations of traditional medicine where, for example, the leaves of *Innula* provide support for the respiratory system.

Let us take a more detailed look at our nutritional relationship with

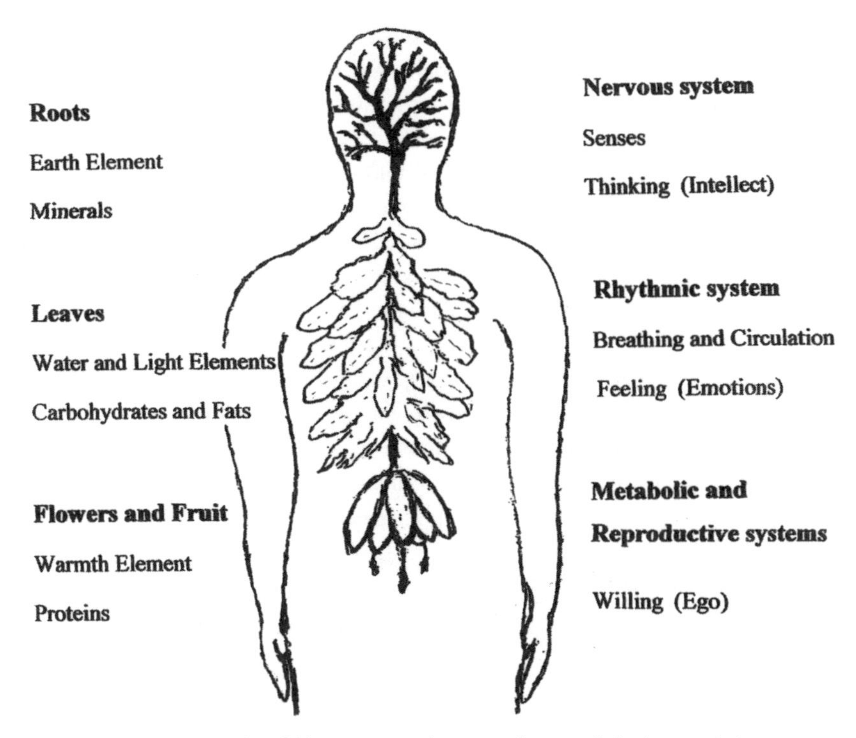

Fig. 10.8 Threefold connections between plants and the human being

roots. Plant roots connect with the earth, with which we associate minerals and mineral salts (ions). The latter we have indicated are of importance for the brain and nerve system. We should realize that it is not dissolved salts as such which are involved—these may be dissolved in water and ingested that way—but their essences or radiations following digestion. These radiations are developed by minerals according to soil condition, the most beneficial circumstances being provided by organic and biodynamic management systems. This influence—of the sun working with the outer planets—was discussed in Chapter 4.

When commenting on the use of human and animal manures in Chapter 3, the notion of 'ego potentiality' was introduced. This residuum of unused forces in *animal* manures can be utilized by the human being.[42] These forces, which may be included with the cosmic influences discussed in Chapter 4, are taken up by the root organs of plants from soil minerals and passed on to us in our food. Minerals that have experienced a recent history in a soil with abundant life quality thus enable the nerve-sense system to be well directed by a person's ego. In this respect, through its negative impact on soil quality, conventional agriculture has a lot to answer for. Thus, through poor nutrition (lack of root vegetables or root vegetables lacking the above forces), those functions directed by the head, *spiritual as well as physical*, may well become compromised.[43]

Such deficiency is likely to generate a cascade of problems, from disorganized nutrition to lack of ability to concentrate and nervous disorders. It may well lead to organs being prone to malfunction and to being less well formed in subsequent generations.

Insight into current health problems

We must reiterate here, the complexity attending all health issues and the need to exercise caution where interpretations are necessarily speculative. With this reservation, we will explore a number of health problems for which insight can be gained from the nutritional principles just outlined. On this basis there is reason to believe a number of conditions could be ameliorated or reduced in frequency, firstly by awareness of environmental risks and secondly by adoption of appropriate diets containing significant amounts of organic and, if possible, biodynamic food.

Fertility

Most couples experience difficulties conceiving at some time or another. The difference today is that this has become an obstacle for at least 20% of couples in the UK.[44] Over the past 60 years the human population has been increasingly subject to synthetic oestrogenic and other endocrine-disrupting chemicals, notably pesticides. In addition, fruit and vegetables have declined in their mineral content while diets have included increased amounts of processed foods with a vast array of additives such as artificial colour, flavourings and preservatives. For a combination of reasons, it appears the body is now less able to raise normal metabolic processes to the level required for reproductive activity.[45] In consequence, since the 1950s sperm counts have fallen by more than half, with sperm quality declining too.[46]

But an environment with raised levels of hormone mimics and disruptors also contributes to miscarriages, to polycystic ovarian syndrome (now affecting 1 in 10 women of childbearing age) and cancers of the breast and reproductive system of both sexes.[47] This must equally be considered a factor in foetal irregularities, leading to birth defects (2–3% of babies) and related social maladjustment. Intake of alien substances occurs by way of food, water, atmosphere and skin. Some substances are persistent in the environment and accumulate in fatty tissue. Our greatest exposure is therefore when eating fatty foods, as well as unpeeled or unwashed vegetables and fruit. Farmers, especially those of developing countries, are subject to direct intake of insecticides, herbicides and fungicides. Substances of industrial origin such as dioxin, phenols and poly-chlorinated biphenyls have been implicated for a long time, while detergents, resins, plasticizers and fumigants have also emerged as harmful contaminants.

At this point we should recall the outcome of animal experiments reviewed earlier. These showed that the starting population was least affected by experimental conditions whereas subsequent generations revealed the true deficiencies of a non-organic diet. This very problem, in the case of animals and their husbandry, was exposed by Rudolf Steiner in the final lecture of his *Agriculture Course*. Thus while 'dead' food and poor living conditions can be managed for a time, the organism becomes weakened, so that *subsequent generations* are less receptive to the normal flow of energies, lack vitality and are more prone to disease.[48] Following the principles stated earlier, such chemical and nutritional provocation could well lead to less strongly formed physical organs and lower vitality. This also helps explain why there has been a parallel increase of early-onset cancers and diabetes.

Cancer susceptibility

Cancer arises from loss of control over cell metabolism and regeneration. Affected cells have been described as cutting themselves off from the normal circulatory process. We have already referred to the body's intake of substances likely to deflect processes from their normal paths—even cosmetics contain these. Meanwhile pesticides suppress the immune system, allowing cancer cells to migrate and cause tumours. The patient is mostly unaware of these various stresses within cells and organs, which may take years to manifest as organic disease. Recently, the density of 'electrosmog' to which our lives are subject has caused concern but, as with chemicals and GM products, the industry's investigations are not uncommonly designed with obfuscation in mind. Nevertheless, WHO *has* classified electromagnetic fields as potentially carcinogenic. This connection seems obvious, for the body's energy or etheric system would seem especially vulnerable.

Connections between diet and cancer are periodically made. The protection afforded by vitamin C and other antioxidants is often repeated. Following previous discussion, *the importance of food with a living quality likely to enhance the body's energy system cannot be overstressed*, together with avoidance of particular foods.[49] However, the ability of the body to coordinate earthly and cosmic nutrition (via the endocrine system) does not depend merely on one's current diet—it is also a matter of the body that has been bequeathed to us! In this respect genetic factors must be acknowledged.

Returning to consider cell processes, research has focused on cations while anions have been relatively overlooked. Iodine, for example, is a substance which, while low in concentration in the body, activates hormones which control heart rate and, crucially, cell metabolism. In fact, iodine plays a crucial part in the aetiology of breast cancer and fibrocystic disease of the breast—possibly, therefore, of other cancers too.[50] In past decades there has, in the West, been an appreciable reduction in iodine intake. The Reference Intake is calculated at a mere 100–150 μg/day. But where levels higher than this occur, as in Japan, breast cancer is less frequent. It would seem that many physicians have a 'thyroid-only' view of our need for iodine.[51] When it is appreciated that any intake of chlorine (water) or fluorine (toothpaste) reduces our absorption of iodine, health authorities should conclude that we require more than the bare reference limit in order to offset risk of deficiency. The practice of filtering water is partially effective but focuses minds on getting rid of what is not wanted rather than putting back what is. Our modern preoccupation with *total* intake rather than ion balance masks an issue with wider ramifications.[52]

It is well established that dietary deficiency leads to illness. That it could be a key factor in cancer formation or suppression is highly significant and underlines the role of the endocrine system—through iodine and the thyroid—in linking the cosmic and earthly interactions of the body. With accumulating evidence of other endocrine glands being affected by foreign substances we should certainly be paying more attention to this issue—an instance where education and preventative measures should prove extremely cost-effective.

Obesity

It is now said that more than a quarter of the UK population is obese. The popular view considers this the result of a combination of diet and drinking habits, a lifestyle dominated by television and computers, and a lack of routine exercise. Given that the vast majority travel by car or other transport rather than walking or cycling, one can understand why these factors represent a cocktail for becoming overweight. While genetic factors may apply in certain instances, these cannot explain the current epidemic. Sleep deprivation is also linked to obesity. Research at Oxford University indicates we now sleep on average two hours less per night compared with the 1960s. It should also be said that low-income families, among whom obesity is most prevalent, are less informed about dietary issues and consume the cheapest available food with a high fat and salt content. Obesity is not simply an inconvenience—it increases exposure to many conditions including coronary heart disease and Type 2 diabetes.

It was pointed out by Hauschka that 'where the cosmic nutritional stream dries up, the organism tries to make up for it by increasing the earthly stream'. Cosmic nutrition is mediated by the endocrine system which, as pointed out, appears increasingly compromised. There is little doubt that much fast or junk food fills the stomach but does not satisfy the body. In consequence a renewed desire to eat arises. In particular, inadequate support is provided for the body's nervous system. The real value of highly processed food—which is not necessarily cheap—is thus exposed for the limited nourishment it provides. Obesity may also be promoted by consumption of homogenized milk and our appetite for animal proteins and fats whose products (according to anthroposophical nutrition) are carried over into the body rather than being fully destroyed by digestion.[53] Further problems arise from substances designed to deceive the body, as with saccharin and an armoury of other artificial flavourings. Many foods and drinks contain the sweetener aspartame. Among the symptoms reported in connection with this substance is a *gain* in weight, for due to its reaction in the body regular consumption encourages a craving for carbohydrates![54]

While the causal factors of obesity have intensified in recent years, parts of the picture have been with us for more than 40 years without such obvious consequences. It was traditionally something associated with older people, but today it is the young who are so evidently affected. This fact suggests that obesity is not simply a result of current fast food and lifestyle—especially as not all are affected—but of a factor making it difficult for increasing numbers to regulate their body processes.[55] As already suggested, youngsters may now have weaker constitutions after 2–3 generations of chemical produce, processed food, and a period in which access to a wide food choice has led to eating selectively—addictively would be nearer the truth—rather than what is nutritionally sound.

All this may therefore underlie a situation where the head and nerve system have gradually lost their capacity to build the body in a balanced way. Indeed, the fact that low-frequency magnetic fields lead to neuro-degeneration merely exacerbates the problem.

Diabetes

Diabetes mellitus is a condition in which body cells are not able to absorb enough glucose sugar—the body's main energy source—from the blood. There are two main forms. In Type 1, the pancreas, an endocrine organ, produces too little insulin or none at all (onset mostly beginning in childhood). In this case, cells have to use fats as an energy source instead of glucose, which leads to a build-up of glucose and toxic by-products in the blood and urine. This form of diabetes requires regular insulin treatment.

In Type 2 diabetes the pancreas produces insulin but, due to failure of insulin receptors, muscle, fat and liver cells are unable to make use of it—they have become insulin-resistant. Ten times more common than Type 1, in the past this affected people over the age of 40 who often developed it gradually, with symptoms going unrecognized for years. Now, diabetes rates are increasing among young people.[56] Insulin resistance is part of a syndrome relating to release of fatty acids, hormones and other factors from the adipose (fatty) tissue of obese persons. These substances, unmetabolized, are then able to interfere with body functions and produce complications. In its early stages, this form of diabetes is normally subject to management by suitable diet and lifestyle.

In the UK over £100m per week is spent on diabetes treatment and by 2006 this had overtaken smoking and related problems. The increase in diabetes in the UK and other western societies parallels an increased incidence of obesity, which doubles the risk of getting the disease. Those affected are also at risk of developing high blood pressure, strokes, heart attack, kidney disease and damage to the eyes and nervous system. Some, including south Asians, are genetically prone to Type 2 diabetes, this applying to Indians and Sri Lankans now living in the UK. A richer diet with higher sugar intake and exercising less are major precipitating factors.

In addition to damage caused by metabolites released from fatty tissue there is the question of substances all of us ingest in food and water—endocrine disruptors occur in many manufactured products. An example being investigated in connection with diabetes and cardiovascular disease is bisphenol-A, which leaches out of plastics into bottled water and other products. But it is pesticides which remain the greatest worldwide threat to the endocrine system thus impeding the harmonious working of earthly and cosmic nutrition. Elevated levels of pesticides have been found in mothers whose two-year-old children already have diabetes, while human breast milk has often been found to contain higher levels of chemical toxins than are permitted in milk sold by dairies![57]

The immune system and hypersensitivity

Immunity is a complex subject and operates on different levels of the organism. Diseases of the immune system (hypersensitivity) are of four types.[58] For example, *excessive immune response*, commonly referred to as allergy, includes asthma, hay fever, eczema and reactions to injected drugs, while other types of hypersensitivity or *autoimmune disease* include polyglandular syndrome, Type 1 diabetes and varieties of anaemia and dermatitis. These are all acknowledged to be products of western society in the past 50 years but are now increasing in developing countries too. So, as if mirroring the endocrine system, the body's immune system appears faltering or unpredictable in its responses.

The immune system, initially from the mother, is something which develops in the course of life.[59] Age and sickness lead to weakened immunity as do courses of antibiotics, radiotherapy and chemotherapy, and above all 'stress'. Stress is a major issue in modern life. We well know how emotional stress affects the digestive system and it is worth focusing on this, for experiencing stress in a conscious way is one thing, but most aspects of digestion and metabolism are carried out unconsciously. For this reason we should give thought to how the body reacts to food additives and artificially reconstituted carbohydrates and proteins. Why, we may ask, are the lives of increasing numbers now blighted by immune disorders?

The issue of nutrient deficiencies should be mentioned as these, singly or in combination, affect aspects of the immune system. Adequate dietary magnesium, selenium, zinc, copper and iron have been identified as crucial.[60] Historical reduction in mineral content of fruit and vegetables, together with increased consumption of processed foods, underlines our vulnerability. Indeed, diets based on refined rather than whole foods may be the root cause of much *food sensitivity*. This is a growing concern and comprises *intolerance* and *allergies*. In the former the body's ability to assimilate particular substances is compromised for some clinical reason, while in the latter the autoimmune system informs the body that something is not safe—gluten for example—and won't be digested. Sensitivity of either type clearly reflects a functional weakness and it is increasingly the young who are affected by these disorders.[61]

As we can see from the preceding discussion, the body is subject to many influences acting in combination. We can therefore only offer the thought that where food intake is consciously monitored and comprises significant amounts of organic and wholefoods, a dysfunctional immune system is less likely to manifest.

Food poisoning and other pathogens

The principal cause of food poisoning is stale or contaminated food. Statistics are problematic but it appears that, while around 40,000 per day die of starvation worldwide, over 15,000 appear to be dying as a result of food poisoning, following its considerably increased incidence worldwide in the past two decades. The pathogens involved vary according to the food source and subsequent contamination. In the UK, while total cases are actually on the decline, the proportion of cases having to be referred to hospital has significantly risen. What we have to consider are the factors leading on the one hand to food becoming stale or contaminated and, on the other, our ability to cope with pathogenic organisms.

Contaminated water is a major cause of food poisoning as well as other serious diseases. There are, however, many reasons for the recent increase of food poisoning, including changing lifestyles and eating habits, a general mobility of people to different regions and, perhaps most importantly, an increased consumption of meat and dairy products. Much food poisoning results from the consumption of raw or undercooked meat and poultry, the reheating of food after storage, or unhygienic preparation.

But methods of farm production are more intensive than ever before, with chickens living less than 50 days before slaughter. In order to achieve this production, they are forced with antibiotics (formerly hormones) beyond their natural growth rates and are made to live in confined spaces—not uncommonly one bird per area the size of an A4 sheet of paper. They suffer ill health and are often barely able to stand.[62] Should we expect that food produced in this way—without feed of real vitality and reared under cruel conditions deprived of normal lighting—keeps as well as meat from more naturally reared animals? One must realize that all food contains bacteria but, as appears evident from tests of keeping quality, their proliferation will be suppressed by higher levels of life energy. Intensive animal production, of pigs as well as chickens, is likely to constitute a further and serious part of the food poisoning equation.

We should now consider *our* ability to resist disease pathogens. Earlier comments suggest our immune systems may often be weak. Contributing to this situation is the passive ingestion of antibiotics used in intensive animal production systems, as well as routine antibiotic prescribing by doctors. On the one hand we have an industrial approach to animal products leading to food which may well be susceptible to rapid putrefaction. This food is then consumed by humans whose understanding of food hygiene is frequently inadequate, whose water supply may be suspect, and whose immune systems are compromised.

But there is a further and alarming dimension to this which transcends food poisoning. The rising number of hospital referrals for food poisoning is partly because standard antibiotics are proving less effective against developing strains of bacteria. This has led to publicity regarding MRSA and *Clostridium difficile* in hospitals. Recently there have been cases of animal-related MRSA[63] along with the more sinister lymphocyte-attacking PVL strain. It is one thing to have problems confined to hospitals and quite another when serious pathogens can be acquired within the community. While bird and swine flu outbreaks rightly receive publicity because of the leap from animals to humans, we should realize that a worst case pandemic could happen—all because of the abuse of animals in factory-farming systems combined with weak immunity and our inability to counter pathogens with effective antibiotics.

Nervous system disorders and ME

Although one may speak about the body's separate systems we are certainly more than just a set of parts! The entire body, together with immune and endocrine systems, cannot function without the support of the nervous system. Diseases specifically linked to the nervous system are also on the increase, one of these being ME. This has been a thorn in the side of the medical profession, for the main characteristic of myalgic encephalomyelitis is chronic fatigue, which, as with post-viral syndrome, results from many prolonged infections. Our interest in this condition arises from the fact that it is a disease of comparatively recent times. Immunization and exposure to pesticide are among a list of precipitating factors, and while allopathic drugs can help manage the condition, progress or even remission may be possible with homoeopathy.[64]

Because ME, or at least its rise in frequency, is a relatively recent phenomenon, one is entitled to link it epidemiologically with modern life, modern dietary factors and medical procedures. It is claimed that the precipitating factor is often a viral infection—ME sufferers frequently complain of muscle pain. Yet we are most prone to virus attack when our immune system is in a weakened state, to which there are various contributing factors—stress among them. This brings us to consider possible connections between immunization and exposure to pesticides. Stress again?

To be clear, the majority of immunizations, MMR for example, are 'active' vaccinations which expose the recipient to an attenuated strain of a virus (or antigen). The latter stimulates an individual to mount a sustained immune response against further exposure to the virus. Thus, from a purely physical standpoint, immunization protects against key diseases. But according to some opinion its successes have been overstated and its risks suppressed.[65] Steiner commented that childhood diseases have a special function in helping us take full possession of the physical body bequeathed to us by heredity. In general they are vital, and to prevent them occurring in the normal way is to risk a certain debility of the organism.[66]

Regarding the link with pesticides, we will consider organophosphate-based insecticides, which British farmers were obliged to use against warble fly, and which were used in the Middle East by the American military.[67] In both cases, those subject to its use have suffered chronic fatigue, as do many farmers throughout the world. Indeed, it was in the First World War that phosphorus-bearing chemicals were used as a *nerve* gas! While direct ingestion is involved in the cases mentioned, pesticides or their residual fragments can persist in the environment and are present in minute doses in many food products.

As it happens, phosphorus has special significance, for it is a 'carrier of light' in our organism and, as such, is crucial for the reception of cosmic energies by the brain (recall reference to phosphorus in Chapter 4). Interference by complex organic forms of phosphate may constitute a significant stress factor to which certain people are more sensitive. Recently a relation has been found between severity of ME, the amount of ATP maintained by cell mitochondria and their rate of carrying out conversions. The latter is influenced by the energy-carrying capacity of phosphate.[68] While beneficial phosphorus radiations are needed from our nutrition, alien organophosphate molecules are likely to be carried over unchanged into the body and have a toxic effect. This connects to the whole controversy surrounding BSE, prions and vCJD. It bears a direct resemblance to trace element toxicity in plant nutrition. It is of interest that homoeopathy has achieved some positive results for ME sufferers. This would appear to illustrate a desperate need for beneficial radiations to be re-established in the body and it supports the nutritional principles discussed earlier—that food substance becomes dematerialized within our organism.

If we recall the connection made between the minerals of root vegetables and the functioning of the head we might perhaps say that this has failed to provide an adequate foundation in many cases. Then, presented with further stress-inducing factors, the brain may lose its cosmic connections, resulting in a person's force of 'will' becoming withdrawn, while in the course of time the brain may be subject to actual degeneration.

Final thoughts

The writer makes no particular claims to authority on either medical or nutritional matters, but what would appear obvious from the above commentary is that society should become much more aware of the *connectedness* between the diseases of our time—otherwise we will never find our way to the root causes, one of which is diet and food quality.

While for many on this planet shortage of food is a worsening prospect, for increasing numbers the food consumed offers little real nourishment and has become a threat to existence.[69] Today, lifestyle and convenience play strongly into what determines food choices. Many simply do not have the time to prepare meals from raw ingredients, while a command of what used to be called 'domestic science' has largely disappeared. Yet home-prepared meals drawn from organic raw ingredients need cost no more than ready meals purchased in supermarkets, and with the benefit of knowing what has been eaten! Of course, family budgets are under pressure and low prices are a major preoccupation. Yet if cost-cutting is the aim, this is a strong incentive for preparing wholesome food, whenever possible, from basic fresh ingredients, whether organic or otherwise.

The truth is, everyone has priorities in the way they spend money. Surprising sums may be spent on holidays, motoring or clothing. Most of us will spend on entertainment, to which eating-out is inevitably linked. And what about the money spent on health insurance premiums? But with food for the home, the attitude is frequently different, so that skimping on the domestic food budget—albeit through ignorance of the consequences—has become a way of life. Television is replete with food programmes suggesting we have a fascination with the subject, though our reasons for watching are as much visual as instructional. It is true that the message about healthy eating is now being heard and out of commercial interest many brands are jumping onto the bandwagon.

But a quantum leap would be made if, through education, it was to be realized that adoption of sensible dietary principles, let alone the consumption of organic and biodynamic food, represented by far the best form of health insurance. The organic dimension might be considered 'a bridge too far', though it is not hard to see that a number of related disorders (part of a syndrome precipitated by obesity) will continue to blight people's health, shorten lives and cripple healthcare budgets unless there is sustained high-profile instruction on nutrition and lifestyle. The latter has recently received media attention, and school meals are now at the forefront of a campaign to improve diet. But a further problem could well be dysfunctional home eating routines, and here while children are always the victims they may also be the prime cause.

In addition to his remarks about nutrition, Steiner explained that vital forces from our food are needed for development of our higher spiritual capacities. He acknowledged this to be a factor reducing people's ability to make strides with spiritual science even in his lifetime. We may conclude from this that our psychological outlook is also affected if food lacks those vital energies and, in addition, contains harmful additives and residues. Lifeless food and a host of alien chemicals have already seriously challenged our physical state and, as time goes on, more and more 'substances' will be implicated in behavioural anomalies, just as with drugs. This should cause us to reflect on the current state of society and levels of alcoholic disorder and violent crime. Of course, many factors lie behind society's problems, but a question worth asking is to what extent a failed nutrition contributes to these.

11. Community Supported Agriculture

By Bernard Jarman

This chapter explores the beginnings of community supported agriculture (CSA) and tackles key philosophical issues relating to the nature of both farming and society. A description given of the 'intentional' Camphill community model shows how it has provided inspiration for the developing CSA movement.

Community supported agriculture as a concept has been around for at least 30 years. My first contact was through Trauger Groh, author of *Farms of Tomorrow*[1] and pioneer of community supported agriculture. I worked as an apprentice at Buschberghof, Fuhlenhagen, the biodynamic farm in northern Germany which he co-founded during the early 1970s. The farm began as a partnership between three farming families. One of them had inherited the property and the others brought capital to it. It was all placed in a Trust whose objectives were to research biodynamic agriculture and its social context. Responsibility for the farm was shared between the three partners; one focused on the dairy herd and creamery, another on cereals and bread, a third on vegetables; overall responsibility was carried collectively. Customers from Hamburg came out to the farm to collect raw milk, quark, butter, bread and vegetables. This was the beginning of what has now become a successfully operating community supported farm. It developed in a region where there were several well-established biodynamic farms and a strong interest in discovering new and less egotistical social and economic structures—both crucial ingredients for developing a community farm capable of withstanding the tests of time.

The nature of a biodynamic farm

It is widely acknowledged that CSA finds particularly fertile soil on farms where biodynamic agriculture is practised. There is good reason for this. Biodynamic agriculture builds on a spiritual understanding of nature and the human being. This means that living processes and their interactions are considered, not just material substances. It also means that each organism, whether plant or animal, is seen as part of a wide and universal whole. Just as Goethe in his *Metamorphosis of Plants* showed that each part of the plant is a microcosm of the whole, so is a biodynamic farm conceived of as an organism and a microcosm of the whole earth. This leads to the fundamental concept of biodynamic agriculture formulated by Rudolf Steiner:

> A farm is true to its essential nature . . . if it is conceived of as a kind of individual entity in itself—a truly self-contained individuality. Every farm should aim to approach this goal as closely as possible—it can't be completely attained—but a self-contained individuality should always be the aim. In practice this means that whatever is required for agricultural production should come from the farm itself, including of course in the farm the due amount of livestock.[2]

Such a 'leading idea' has enormous consequences for a farm operation because it encourages economic and ecological self-sufficiency. A strong internal cycle develops from soil to fodder to livestock and back to soil, as discussed in Chapter 3. This builds fertility and the stronger this cycle becomes the more is a farm able to produce surplus crops which can be sold off the farm without depleting fertility. It is this internal cycle which gives the farm individuality its strength and, ultimately, its 'immune resistance'. The grass, the cows, their intestinal flora, the soil micro-organisms, adapt themselves ever more strongly to one another and gradually over time an individual identity begins to emerge that is unique and site-specific. It becomes in effect a living being with a biography (see also Chapter 9).

Caring for this living farm organism and making it productive is very different from an industrial process. Farming cannot be described as an industry since everything that is produced arises as a by-product of the farm's internal cycle. Industrial products are created by transforming raw materials through a process, which necessarily reduces them quantitatively. So, for example, a large amount of wood must be cut away from a tree trunk and disposed of in order to create a finely crafted chair. Farm produce is very different. It is the result of a living process. One seed is planted and hundreds may be harvested from that one plant.

In this way the farmer cares for the whole farm by nurturing this individuality and enabling nature to offer the produce as a gift to humanity. This leads directly to the ideas behind community supported agriculture. *A farm conceived as a living being is the foundation for a true community around a farm.* Just as a healthy farm evolves out of its own resources, so can its 'surplus' provide for a community. The more diverse a farm becomes, the more it can fulfil this requirement—for example by providing fruit, herbs, eggs and even textiles as well as milk, vegetables and meat. Diversity benefits both farm and community.

Egoism or altruism in human relationships?

On the face of it, this sounds like self-sufficiency extended to a group of people. There is, however, a fundamental difference. The goal of self-sufficiency is to become independent of one's fellows. Robinson Crusoe surviving on a desert island and providing for all his *own* needs is an idealized picture of this. It is a powerful image and one that underlies much of the thinking behind contemporary economics. It is taken further in Ayn Rand's novel *Atlas Shrugged*. Woven into this story is her belief in the primacy of individualism, independence and capitalism—*I will serve no man and let no man serve me* is the motto formulated, and it found resonance particularly among those seeking to justify the free market in the face of encroaching state control and interference. Like the story of Robinson Crusoe, it provides a sense of freedom, individual independence and self-worth. Its fundamental flaw is that it is entirely self-centred and without compassion. It is this very thing that can make capitalism so destructive. A key economic driver identified by Adam Smith in his *Wealth of Nations* was that of self-interest and competition, as well as the concept of the 'invisible hand' which anonymously regulates supply and demand. It provides freedom to innovate, grow and develop but only by exploiting others and robbing the earth. A parallel is found in the biological sciences. Beginning with the idea of evolution as expressed in Darwin's *Origin of Species*, a parallel theory identifies evolution's driving force as 'the survival of the fittest'.[3]

Another and quite different understanding is steadily gaining acceptance today. It is one that relates closely to the holistic principles behind biodynamic farming. Instead of competition being the driving force of evolution, a new ecological understanding of the world has been emerging. It came to prominence in 1970 with the publication by James Lovelock of *The Gaia Hypothesis*. He suggested that the whole earth is self-regulating and that all life forms act together with their environments within a single homeostatic system. This implies that each species exists not for its own sake but for the mutual benefit of the planet. Although a relatively mechanistic concept, this goes a long way towards recognizing the interdependence of life and challenges competition as being the sole driver of change. Just as the accepted theory of evolution appears to have translated into our current western economic system, attempts are now being made to replace this self-seeking economy with one based on mutual support and cooperation—a Gaian economy. This is the context within which CSAs belong. Instead of seeking to maximize personal gain, the gesture is one of service to humanity.

In keeping with this principle it can be readily understood that every person on this planet has a birthright to sufficient food, clothing and shelter to maintain him/her throughout life even with our ever-expanding human population. An anonymous proverb probably coined at the time of the enclosures says: 'There was a time e'er England's griefs began, when every rood of land maintained its man'. A rood was a quarter of an acre. If today the total cultivable area of Britain is divided by the total population, about one third of an acre is available per person. Naturally there are great regional variations in climate and soil fertility but this calculation provides a useful guideline for anyone wishing to provide members of a community supported farm with everything they need. Thus 100 acres (40 ha) of average English farmland should be able to provide for the needs of up to 300 people.[4]

Background and principles of the Camphill movement

The model for CSAs developed by Trauger Groh was partly inspired by experiences gained from the Camphill movement.[5] Camphill is recognized for its work with people with special needs and for its innovative community life. It was founded in Scotland by Karl König, a far-sighted pioneering doctor who was forced to flee Nazi-controlled Austria in the late 1930s. König used his medical knowledge, the inspiration of Rudolf Steiner and extensive understanding of European cultural history to develop his ideas for a healthy community life. He recognized three semi-autonomous functional systems operating within the human organism:

- The nerve-sense system extending throughout the body and having its main focus in the head, where the brain and most of the five senses are located.
- The breathing and circulatory system, centred in the heart and lungs and which ensures that oxygen is transported throughout the organism via the bloodstream and exchanged with carbon dioxide.
- The metabolic system in which food is digested and transformed, the primary location being in the abdomen.

These three systems—nerves, circulation and metabolism—exist in a dynamic balance with one another. Each one has a degree of autonomy and yet acts in harmony with, and complements, the other two.

During the period of global meltdown immediately after the First World War, Rudolf Steiner tried in every possible way to encourage those in positions of responsibility to develop a more enlightened approach to human social organization.[6] He proposed that in place of the failed concept of a unitary nation state, in which human rights, individual initiative and economic interests continually collide and undermine one another, society should be reordered so that these elements have space to fulfil their true roles. Just as a farm can be conceived as a living, integrated organism formed as an image of the universe so too can a social organism be developed as an image of the threefold nature of the human being. This has become known as the 'threefold social order'. Like the human organism, the body social is thus differentiated into three distinct functional areas:

- *Cultural and spiritual life*. This includes all individual initiatives which thrive when ideas can unfold freely: education, medicine, agriculture and the arts.
- *Rights and the laws of society*. These should apply equally to everyone. This is the sphere of the democratic state.
- *Economy*, namely how we provide for, and meet, one another's needs through production, distribution and consumption of commodities and services.

The principle for the cultural sphere, and for taking individual initiatives, is *freedom*. The principle for the rights sphere, the state, in which everyone should be equal before the law, is *equality*. In the economic sphere the guiding principle should be *brotherhood*. Freedom, equality and brotherhood are three great ideals of a utopian society, and were those of the French Revolution. But they are also very practical, appropriate and no longer utopian when applied to the three functional aspects of social life.

The three principles can also be harmful if exercised out of context, as for instance when the economy allows free reign to individual egoism rather than providing a brotherly service to the community, or when equality and uniformity places individual expression in a straitjacket, or again, when the state becomes a brotherhood of interest groups instead of promoting justice and equality. Careful differentiation between these three parts of society is essential and is acknowledged as a reality in today's world. Thus it is essential for consultation to involve businesses, governments and non-governmental organizations. These three groups each have a distinct purpose and are representative of the three spheres of society—economy, rights and culture respectively.

In setting up Camphill, König worked with these ideas in a unique and practical way. Besides the guidance of Rudolf Steiner he was influenced by the ideals and achievements of great social reformers such as Amos Comenius (1592–1670), Ludwig Zinzendorf (1700–60) and the Welsh social reformer and industrialist Robert Owen (1771–1858). In his essay *The Camphill Movement*, Karl König describes how these personalities, in successive centuries, developed different community-building qualities which have had a lasting effect on European culture. Comenius brought the idea of universal learning as a cohesive social force. It led to the founding of Europe's great universities and learned societies. Zinzendorf created Christian-based communities that welcomed all-comers equally, regardless of their faith or status. Owen brought the idea of an economy based on brotherhood and mutuality. This led to the founding of the Co-operative Movement whose first model community, as with the first Camphill, was established in Scotland.[7] To these we may also add the contribution of William Morris (1834–96), English poet and socialist thinker who was a pioneer of the Arts and Crafts movement. The latter has become an essential element of the provision of meaningful work within Camphill communities.

König also sought to disentangle unhealthy elements inherited from the unitary state concept, notably where land and labour became a commodity. His solution to the question of labour in particular became the hallmark of Camphill. No one is paid for their labour. Instead everyone works to serve the needs of the community as a volunteer. The community in turn provides everyone with their daily requirements. No one receives wages (see below) nor are there any personal financial incentives. Work is undertaken out of love for the task in hand. The Camphill system is a very specific interpretation of what Rudolf Steiner formulated as the fundamental social law:

> In a community of human beings working together, the well-being of the community will be the greater, the less the individual claims the proceeds of the work he has done himself; i.e. the more of these proceeds he makes over to his fellow workers and the more his own requirements are satisfied not out of his own work done but out of work done by others. Every institution in a community of human beings that is contrary to this law will inevitably engender in some part of it, after a while, suffering and want. It is a fundamental law which holds good for all social life with the same absoluteness and necessity as any law of nature within a particular field of natural causation.

How then does the farm fit into all this? Farmers work with nature. They learn to understand and recognize her imperatives. They want to produce food. They care for the land, the animals and the plants, and then if they are successful nature provides the goods. This, as explained above, is different from any industry, and hence it is wrong to speak of farming as an industry. Farmers are artists and craftsmen and have a role not dissimilar to that of the conductor of an orchestra. The farmer needs to know the right moment to plough a field, graze a pasture or sow seeds. He needs to sense when to engage each aspect of the farm individuality so as to achieve a harmony of life. Farming rightfully belongs to the sphere of free cultural activity. It is not in the first place an economic activity, although those who farm, still need to make a living. The reality is that so long as the crops are in the ground they are not yet commodities. Grains, vegetables or milk only become commodities and enter the economic domain once they have been harvested. Before that they are bound up as part of nature.

Many Camphill communities have sought to reflect this in the way they handle produce from the land. Land payment systems have been set up whereby the community makes a collective payment to cover the external costs of running the farm (purchases, insurance, etc.) but not the vegetables or milk produced. These are received as gifts from the farm and are distributed freely. Within an 'intentional' community like Camphill an opportunity clearly exists for allowing a farm to function in this way.[8] Trauger Groh and others round him then asked whether such a system could be adapted and applied to a more loosely connected group of people living independent lives rather than in an enclosed community. In this way the idea of community supported agriculture was born.

Community supported agriculture—a theme with variations

How community supported farms have been created varies considerably, as does the way they budget their costs. In the North American model the following procedure is common. First a detailed budget is put together by the farm group. Once budgeting has been done, a meeting is arranged for all farm community members. This usually takes place over a whole day and is like a little festival. The farm year is reviewed and plans for the coming season are shared. The farm management group then presents the budget. Every detail can be questioned. At the end of the process the members agree the budget. Having done so, they must find a way of meeting all the agreed costs through their collective contributions. This process is best described as 'auctioning the budget'. On the basis of a rough guideline (budget divided by members) each person pledges to pay a certain sum within the framework of what is individually affordable. There are no fixed pledges (shares). This makes it possible for those on low incomes to pay less and, more importantly, allows those who can to pay more. If on the first round the total pledges do not cover the budget, a further round of pledges is made until the budget is met. Expenditure may also be reduced. This system works well as the commitment is for one year only. Payments are then usually made by monthly standing order. In the USA, where the movement has gained considerable momentum and where there are now over 1000 CSA farms, a specific form has been adopted along the above lines, although in many cases confined to vegetables. In regions with a long winter, this has meant closing down for about four months.

In the UK, until recently no functioning examples of the above kind were to be found. Here, the concept of CSA covers a spectrum ranging from local food initiatives such as farmers markets and box schemes to farms supported by surrounding communities, to 'intentional communities' such as Camphill. CSA is thus defined as a partnership between farmers and consumers, where the responsibilities and rewards of farming are shared. This broad spectrum of local food initiatives was recently recognized in the Soil Association sponsored project 'Cultivating Communities'.[9]

A British example of a full community supported farm is Stroud Community Agriculture (SCA), a biodynamic farm close to Stroud, in Gloucestershire.[10] From a modest beginning in a walled garden it grew during the following six years into a farm with around 48 acres located on two different sites. It is structured as a cooperative (Industrial Provident Society), now with 180 members (200 is the maximum). Each member has one vote, pays £24 per year, receives a quarterly newsletter and can register for a vegetable share. Currently (2009) this costs £33. Payment is made monthly by standing order. There is a set allocation of vegetables per share each week, which is collected either from the farm or from drop-off points in nearby towns. Members also have access to meat, which can be purchased from the farm. Regular farm days take place each month and members can join in activities, help on the farm and have fun. A number of objectives were formulated early on and are written into a constitution. These are to:

- Support organic and biodynamic agriculture
- Pioneer a new economic model based on mutual benefit and shared risk ensuring the farmers have a decent livelihood
- Be fully inclusive—low income shall not exclude anyone
- Encourage practical involvement on all levels
- Be transparent in all affairs; make decisions by consensus and strive for social justice

- Offer opportunities for learning, therapy and reconnecting with the life of the earth
- Network with others to promote community supported agriculture and to share experience
- In cooperation with the farmers, encourage members to use the farm for individual and social activities and celebrations
- Develop a non-exclusive sense of community round the farm

Operating on a different basis is Tablehurst and Plaw Hatch Community Farms in Sussex.[11] Here, all produce is purchased through farm shops and members are not required to carry the running costs by subscription. Yet it is a vibrant farm community with considerable outreach. Faced with the threat of closure in the early 1990s, a large number of individual supporters rallied round, formed a co-op and raised funds to purchase the assets of Tablehurst Farm. This was facilitated at the time by issuing £100 shares. The land and buildings were subsequently transferred to St Anthony's Trust which already owned Plaw Hatch Farm. The co-op is made up of local supporters and customers and, as with the Stroud Community, the members elect a carrying group which decides overall policy on their behalf. The farm teams carry out day-to-day management of the farm within a 'company' which is wholly owned by the co-op. Supporters continue to help raise funds for farm development by purchasing further shares while the co-op also attracts larger donations.

In tracing the origins of such developments in different situations, one observes that initiatives must be met by a receptive human environment. Thus, the Tablehurst and Plaw Hatch development undoubtedly had its precursor in the 1970s when the Anderson family ran Busses Farm as a social enterprise, leading in turn to the establishment of The Seasons cooperative shop in Forest Row. Thus a generation who had experienced the success of former initiatives became willing and committed participants in a later vision for a community farm venture.

In a similar way Stroud Community Agriculture has had a long 'prenatal' history. The two estates currently managed by SCA belong to Hawkwood College and Wynstones School, both strongly connected to one another since the 1940s. The site at Brookthorpe (near Wynstones) used to be run as a productive smallholding serving the needs of Wynstones hostel. The first attempt at starting a CSA occurred in the 1980s. Oaklands Park, a Camphill initiative on the opposite side of the river Severn, had recently been established and, with it, a large horticultural operation. Since the main interest in biodynamic vegetables at that time came from Wynstones, it seemed natural for Oaklands and Wynstones to link up to supply vegetables. They sought to pioneer a land payment system similar to that operating in Camphills. While this original idea had to be abandoned, a second attempt in the 1990s established Kolisko Farm on the Brookthorpe site. Its objectives were to create a biodynamic farm that could integrate agriculture, medicine and education as espoused by the Koliskos in their book *Agriculture of Tomorrow*. After some ten years it became necessary to dissolve this operation, but with these seeds the present SCA was able to start growing.

Securing land for biodynamic farms

Both land and labour are treated today as if they were commodities to be bought and sold on a par with products and services. According to Rudolf Steiner and many other thinkers too, this distorts the economy and causes a lot of misery. Buying and selling *human beings* constitutes slavery and is no longer permitted. Buying and selling *human labour* involves trading part of a human being and is therefore partial slavery. Similarly, freehold rights and the possibility to buy and sell land as a commodity for industrial and housing development create unreal and inflated prices, leading to a serious distortion of the agricultural economy and rendering land inaccessible to young farmers. Opportunities for speculation likewise mean that there are fewer farm tenancies available.

The way forward must therefore be to find ways of freeing the land and making it available for those who can, and wish, to work it. Farming is for the long term and it is essential that farmers are granted a lifelong security of tenure, or for as long as they feel able to farm. It is also vital that land can be held in perpetuity for agricultural purposes, and even that it should be managed biodynamically—a scenario which of course also presupposes the availability of sufficient biodynamic farmers. The example of Tablehurst and Plaw Hatch Community Farm offers a good example of how an educational trust (St Anthony's in this case) can hold land in perpetuity for the benefit of biodynamic agriculture. There are other similar trusts operating across the country. Almost without exception these are educational trusts and there is always the danger that educational objectives override those of agriculture. This is most clearly seen where an educational establishment owns the land but lacks sensitivity towards the farming. Rare are charitable trusts established with agriculture as the primary purpose—indeed charity law makes this hard to justify.

A Community Land Trust (CLT) is another legal structure for holding land. It differs in that it is a co-operative rather than a charitable trust. The best known example of a CLT is that of Fordhall Farm, an organic farm that was purchased by a large number of supporters each of whom bought a stake in the farm in the form of shares.[12] One advantage of a CLT over a charitable trust is that the local community guarantees its continuity rather than a group of nominated trustees. Unlike an educational trust, it is not bound by an educational objective. Its purpose is simply to hold land in trust for as long as required by the community and its different stakeholders. Such community ownership also helps to build local ties and weave greater identification with the land.

Whether land is held in a CLT or in a more traditional charitable trust, an important objective is fulfilled, namely that the land can no longer be used as a speculative asset. Arriving at the point where land can be held in a neutral form of ownership structure requires either that land is gifted to a trust or is purchased on the open market with a large amount of gifted capital. The Biodynamic Land Fund[13] has recently been established to raise funds in order to purchase such land and hold it in perpetuity for the benefit of biodynamic agriculture.

A concluding note

More and more people are realizing that it is no longer responsible either for human health or the environment to rely on chemically grown food, or food which has been imported at huge cost in terms of energy and livelihoods. Testimony to this growing awareness is the growth of the Transition Town movement in which local people have come together in order to pioneer (or return to) more local and environmentally sustainable lifestyles.[14] Its purpose is to make a transition from profligate consumerism, requiring constant economic growth, to a stable and self-reliant economy. It is interesting to reflect that, although quite 'new' for modern culture, historically this was the way the vast majority of non-urban dwellers interacted with each other for centuries.

Community supported agriculture, however composed, is a most effective way to re-localize our food economy and address key issues such as that of food poverty. Although individual share payments to an average CSA may be higher than simple box schemes, in the long run it is one of the most cost-effective ways of obtaining fresh biodynamic food. Comparisons between a typical CSA share and a similar range of produce purchased in a supermarket repeatedly show CSA shares as offering value for money.

It is also not just the price of specific produce that affects or ameliorates food poverty but the way we relate to and source our food. Using fresh, seasonal farm produce requires a change in attitude and cooking habits. A reduction in the use of tinned, frozen or processed food (which is also expensive) in favour of what can be obtained from the farm, benefits not only personal health and the environment but also the pocket. Care of the earth, healthy eating and good housekeeping go hand in hand (see Chapter 12). By participating in community supported farm initiatives, real and lasting social changes can be encouraged.

12. Looking to the Future

Confronting a changing world

At the present time we are literally able to look two ways—back at how things were, and forward at how things will have to be. We are being forced by circumstances, if not by moral imperative, to face a rather different future. The world has always been a changing place but as regards environment and lifestyle we are now living on a cusp.

While global warming and the world economy appear to override everything, wider environmental concerns demand responses individually and collectively on every scale, local, national and international. In this respect, what were originally considered parts of activist agendas have now reached critical mass and are accepted by the majority. As always, the issues are how much can be achieved individually and what leadership we expect from local and national governments. The individual has a certain scope to alter their patterns of consumption and contribute to recycling. But we are also trapped by the way in which our communities are laid out, in particular by our dependency on personal transport, with sizeable chunks of energy consumption committed simply to pursuing daily life.[1] Local and community responses to these problems will therefore become increasingly important and we will need to develop realistic strategies.

An ever-widening catalogue of problems is being encountered. Greatly increased costs of energy and resources have been focusing minds—more so, the increased costs of food as world surplus becomes a thing of the past and as harvests prove more unpredictable.[2] The prophets of old might have seen signs in the sky foretelling crises—we, for our part, have climate modelling and planetary-wide media coverage. Food shortage and famine are now likely to stalk our planet while deterioration of water supplies and water quality in many regions makes the spectre of widespread disease epidemics and food poisoning the more likely, together with conflicts over access to water. Disease-carrying organisms—subject to migration and mutation—will present new challenges both for health services and agriculture.[3] Rising sea levels will demand enormous and ultimately futile expenditure to protect infrastructure and livelihoods all around the globe. Circumstances such as these cannot avoid impacting on the well-being of people everywhere and will increase the pressures on governments, especially those of a democratic complexion.

The environmental crisis is a wake-up call, for questions are being asked about all forms of consumption and waste. It is truly a human crisis for it causes us to question the materialistic and technology-dependent lifestyles to which we have grown accustomed. While the latter allow us to be highly productive, and in obvious ways increase human contact, electronic devices for workplace and entertainment, along with artificial radiations which encircle our lives, are contributing to the isolation of human beings from their cosmic environment. It is significant that different levels of consumer culture now extend to all areas of the globe. This means, broadly speaking, that the world community is confronted by a similar range of questions. Human beings have to this point been offered a great deal by nature,[4] and now the future of the earth is in our hands. Self-realization of this kind will be fundamental to living in tune with future conditions rather than fighting to preserve the past. Those with wealth may feel able to ignore these issues, but in due course they will impact on all. Such philosophical questions are, of course, bypassed when whole populations are subject to cyclone, earthquake or famine. Raw suffering is the pervasive experience when lives are reduced to the most basic struggle for survival.

One key aspect of those 'future conditions' just referred to relates to climate. Despite the apparent consensus on the subject of global warming there is, in truth, much less certainty about what the future holds and even some evidence of pressure being placed on scientists not to 'rock the boat'.[5] A recent trend, linked to minimal solar activity, geomagnetic characteristics and consequent cloud cover, suggests that cooling is on the agenda! While global warming has had highly publicized consequences, cooling would arguably be worse for agriculture and domestic life at a time of food shortage and increasing energy costs.

But whichever becomes the dominant trend, agriculture in the coming years is likely to be plagued by unpredictability and extremes of weather—a situation where the lessons learnt one year are rarely applicable to the next. Surely organic production systems can respond to this situation. They are resilient because they are founded on nurturing soil fertility and promoting a more natural resistance in plants. They promote a healthy soil life, store more carbon than conventional systems and are more sensitive to cosmic influences.[6] They are also manifestly superior for resisting drought. In these respects the productivity and practicality of organic agriculture need to be re-evaluated and misconceptions set aside. And because organic systems do not require synthetic chemicals and deploy local resources, they consume significantly less energy. As energy costs become an increasing preoccupation, this is a further basis for reappraisal of organic agriculture by those who have formerly dismissed it.[7]

The organic and wider agro-ecological movement will need to coordinate their efforts, for what needs to be considered is a thoroughgoing reappraisal of how agriculture is designed on the ground. Depending on the severity of circumstances and the drive towards localized production systems, ideas drawn from agro-forestry and multiple cropping will require consideration. The latter contribute to local needs as well as to concern for the global atmosphere. Growing trees allows carbon to be stored while affording shade, shelter, soil conservation and even green manure. Multiple cropping offers significant productive advantages while responding to local needs.[8]

Meanwhile, the wider development of organic and eco-agriculture is underscored by biodynamics, which validates and reinforces a commitment to biodiversity through a deeper understanding of life processes. In addition to the 'traditional' biodynamic measures, a new generation of potentized preparations—increasingly tried and tested in the field—offers a prospect of alleviating a wide range of problems, including pests, pollution and climatic stresses. Within the biodynamic movement there is also experience of operating on a community-basis rather than simply in a commercial manner. As economic conditions increasingly act as drivers for local production and consumption networks, community agriculture in varied forms seems set to gain momentum.[9]

Counting the cost of food

Confronted with recent rises in food costs, consumers should be aware that in the early 2000s the average British home was spending around 10% of its disposable income on food, the lowest proportion in Europe, and a drop of some 15% since the 1960s. As the cost of food will remain a critical issue, particularly in relation to organic products, it is pertinent to make three key points about it that also help explain the above figures. The first is that to improve food security, western countries from the 1950s provided their farmers with a price support mechanism together with capital grants for farm development. A source of great bitterness in developing countries, this not only achieved but exceeded its purpose while setting the seal on a food culture in which the consumer was protected from the true cost of production.

The second point is that most of this production came about through a chemically based agriculture which, while efficient and profitable from an industrial point of view, was damaging to the environment. In short, part of its true costs were externalized so that while soils and environment suffered residual costs such as those of water treatment and healthcare could be disguised in people's utility bills and private expenditure. Thirdly, with a developing global economy, a widening range of exotic food items have been sourced from wherever production costs are least. While promoting commercial activity in many poorer countries, much of this produce has only the slenderest claims to fair trade. It has been damaging to the environment and water resources of local people, leaving them with fewer options on which to resume their lives were market conditions suddenly to change. So, quite apart from having spawned a generation with little notion of seasonal foods, we fail to pay an ethical price for the food we eat, some of the profit going to offset the absurd transportation costs. Moreover, through our preoccupation with cheap food—notably meat products—we perpetuate practices that are perilously close to causing a major bacterial epidemic for which there is currently little defence.

It was commented in Chapter 10 that cooking from basic ingredients could reduce the burden on the domestic budget if circumstances allow. It has also been estimated that up to one third of the food we buy may be wasted. Wastage is the result of fear and ignorance, as well as the fact that refrigeration allows us to store far more than we can quickly consume.[10] Clearly, were our purchasing behaviour to be more disciplined, the money saved might go towards food which many claim they cannot afford. Quite apart from this, organic produce contributes to reduced food wastage through remaining fresh for longer.

Current pressures on the household budget inevitably impact on organics, so it is well to consider carefully the foregoing remarks. Organic food may cost more but could lead to an improvement in general health. Whether more of the public can be persuaded to eat organic as the best form of health insurance remains to be seen. It is also said that organic food is more sustaining, so it might well be the case that we should consume less but of better quality. Although more research needs to be conducted into the efficacy of organic diets, such is the enormous scale of expenditure by the NHS and other healthcare systems that huge savings would be likely to result if diets were to include fresh food produced by organic or biodynamic means. This underlines the failure of a policy which until recently has emphasized *sickness care* rather than *promoting good health*.

But as many would concede, buying organic food is not simply about one's health; it supports as far as possible environmentally sound agriculture. To an increasing extent, it also supports farming in the locality. It is a tragedy that the elevated prices of organic food items reflect a speculative attitude of retailers rather than a generous return for organic producers. This is a strong argument for community marketing schemes capable of supporting producers and protecting the environment *without* alienating an increasingly concerned public.

How secure is British food?

The question of food security offers different pictures when considered within national or global contexts. Except during wartime, farming in Britain has never fully been geared to national needs. First we relied heavily on the Empire, so that joining the European Union consolidated rather than altered this reality. But sadly, as British farming has proved increasingly uncompetitive, generations of farming expertise have effectively been sacrificed on the assumption that cheap food could be imported in perpetuity.[11] Such a policy, unmatched in other countries, always carried the seeds of a national food crisis. It has chosen to ignore the probability that at some stage world surpluses would be eliminated and assumed it would always be more cost-effective to import a major part of our food.[12] Through such a thoroughly distorted trading economy, many of our British farmers are now little more than park keepers while their average age verges on that of retirement!

This betrays the essentially short-term expedients which have characterized British politics, the idea of maintaining agricultural capacity having been no more attractive than improving public transport. In view of the currently worsening world food situation and the cost of fuel, sharp corrections of agricultural, marketing and distribution policy are surely required in order to create more local networks. On a personal level, this has already led to increased garden vegetable growing and a resurgence of interest in urban allotments.[13] Indeed, there are now more instances of fields being made available by landowners for local vegetable growing. It is into this scene, concentrating on the use of local resources, that organic and biodynamic agriculture must find a renewed sense of purpose.

The further dimension of the British problem is that since only about 1.5% of the working population are engaged in agriculture there has been little recent attraction to a future in agriculture, and since the whole educational system is aimed at other objectives, young people—now almost entirely from urban backgrounds—can hardly be expected to see a role for themselves in food production. Furthermore, a sizeable number of those employed in agriculture are migrants who are either prepared to do jobs, or to work for wages, unacceptable to British people. Again, land values are so high that young people have little hope of ever owning a farm, and yet without youth and commitment to agriculture our future is dire. It is interesting to note that a number of those now entering farming do so either in parallel with or following a career elsewhere.

A global view of food security

On a global scale, the burden on world agriculture is being made ever worse by an increasing population. In this respect it is time to stop idle discussions about how we plan for 9 billion by the year 2050 and instead look to how we can stabilize the global population in order to preserve the planet and at least a limited quality of life. Such an attitude may appear Malthusian or in some other way politically incorrect, but it recognizes that we cannot tackle the existing problems of the planet if we take our eyes off one essential overriding factor. An imperfect metaphor perhaps, but it is like trying to fill a bucket with a hole in it! Meanwhile an image of modern food security, which was drawn some years ago, is that of the pyramid. In ancient times, such as those of the Pharaohs of Egypt, the proportion of the population engaged in food procurement could be likened to the pyramid's base. Nowadays, those in agriculture would probably represent the apex. How well does a pyramid balance on its apex? We are in the process of finding out.

In the tropics and subtropics, in addition to areas lost to desertification or salinization, soil exhaustion has rendered once-productive land uneconomic for cropping, notwithstanding relatively low labour costs. More fertilizer and pesticide is required just to maintain yields above an economic threshold. Many farmers unable to match costs against yields have committed suicide. This underlines the false promises of the Green Revolution, for many such areas are supposed to have benefited from high-yielding varieties. Now, especially for the African continent where drought is a perennial problem, GM crops are being actively promoted as the big hope for future food and prosperity. Have we not been here before? Many of those who don't necessarily support GM nonetheless display an abiding faith in technology as a way out of future difficulties rather than adopting the simple credo of living in a more balanced way with our environment.

Experience has already shown that transgenic crops are failing to achieve promised yields, and there is no basis for a belief that such crops will escape diseases and pests for which they are *not* inoculated. In the meantime they have contaminated indigenous varieties and have slipped through without rigorous tests of food safety (see Chapter 7). As such, they should be regarded as a weapon of biological pollution and potential mass health deterioration.[14] Research on their efficacy has been subject to corporate secrecy and only its most promising results have been publicized. One can understand that destitute people might welcome any solution, yet GM crops now threaten Africa and other regions with continued rural poverty and enslavement to major corporations without any guarantees of food security.

There is a prevailing view that 'only by using chemicals and high-tech farming will we manage to feed the world', although throughout much of the last century organic methods haven't been given a chance. And one can, of course, adopt the view that chemical farming, in forcing our world to grow more at any cost, has fuelled the population problem we now face. Yet organic systems can and do outperform conventional methods in low rainfall years and, in the modern context, have largely unexplored potential in the various forms of multiple cropping. Organics also perform well under most tropical circumstances.[15] This is attributable to low inherent soil fertility, high temperatures and intense rainfall characteristics. This combination of factors highlights the fact that the export of chemical agriculture to these regions constituted a totally inappropriate transfer of technology. However, on account of the inefficiency of chemical farming it has proved an ideal business model for the multinationals.

It is fair to say that many who support high-tech approaches are driven to greater or lesser degree by corporate interests or by research funding connected to the latter. They are prepared to ignore the reality of widespread soil deterioration or the attendant human collateral, in order to preserve the status quo. Chemicals and transgenic crops are in reality a burden the world carries rather than a solution to its food needs. Indeed, as other authors have pointed out, a great deal of crop production is destined for trade in agricultural commodities rather than to feed local people. To claim that high-input farming 'feeds the world' is therefore a spurious argument. Current realities suggest that we will be unable by these means to maintain present levels of production, let alone accommodate future increases.

On the other hand reversals of agricultural fortunes have been achieved with organic and biodynamic methods in many parts of the world on soils which were in a degraded state, such as tea-growing areas in Sri Lanka and cotton-growing areas in India. Notable examples also include the use of legume crops such as *Mucuna* beans in Latin America,[16] there being hope this may be the way forward in Africa too. In general, the potential of organic agriculture has been overlooked because, unlike chemical farming, it has lacked energetic promotion, research backing or effective extension services. In particular, the capability of biodynamic practices to confront poor or degraded soils needs to be more widely recognized. So bearing in mind the need to take extra care for the earth's remaining resources, a rational rather than commercial assessment of global food security would lean heavily towards organic-based farming systems rather than a dependency on high-tech intervention. This view is further endorsed when it is realized that by this means the *food sovereignty* of local peoples can be maintained and developed.

Challenges for the organic movement

From the foregoing we should expect organic methods to be making a major contribution to the future of world agriculture.[17] But will this happen?

Institutional and commercial constraints

The prospects for organics and other 'alternative' movements would be brighter were it not for the desire of major commercial interests and their supporting institutions to manipulate and control the world to further their own interests and to constrain the freedom of others, culturally and economically. Superficially we have democracy, but the realities of economic life are largely ruled by unelected and invisible players. This is a case where the economic sphere has overstepped its true function.[18] A prime example is the World Trade Organization (WTO). Driven by transnational corporate interests, it has been the major influence on globalized trade, facilitating the activities of major players. But WTO has powerful influence with the UN, WHO and FAO, so that bodies like the European Community, let alone national governments, are unable to curb these powerful multinational forces. One result is that under the banner of harmonization and consumer protection FAO's *Codex Alimentarius* (Food Code) will exercise increasing control over global food and health. Even organic farming is within its sights.

The principle of democracy, however interpreted, is preferable to this control by stealth. Here and elsewhere, the slow drip of increased regulation is in danger of bringing to an end or tethering individualism, along with freedom to use traditional or 'alternative' practices.[19] The latter are no burden on national economies, offer low cost solutions in a cash-strapped world and could be contributing so much more. The steady rise of organics, together with alternative medicines and healthcare, has clearly been a concern to the more powerful institutional and commercial forces. This is evidenced by their continuing need to dispute the quality of organic food or to threaten the efficacy and legal status of homoeopathy and other alternative therapies.

We should reflect that while we are increasingly crushed by bureaucracy, a high proportion of innovation is brought about by individuals and small enterprises. Regulation not only imposes unacceptably high set-up costs, but subsequently higher overheads which have to be borne by consumer and taxpayer. These costs deter investment while restricting economic viability. A previous generation murmured about the impact of mechanization on jobs; ours has been pitifully slow to realize that people are put out of work by forces which individual governments have been unable to control. Looked at more widely, as economic conditions become more challenging, over-regulation seems likely to precipitate the nemesis of the economic system quite as effectively as any failure of credit.

Meanwhile the overriding objective of powerful interests is to preserve the status quo. Despite the consequences for environment and human beings, major companies will seek to extend their dominion.[20] It would seem that the same principles apply here as for the survival of species. Many people are inclined to say 'the major players are here to stay so we might as well put up with it'. But maintaining 'business-as-usual' will end up exhausting the world's land resources and, in the case of chemicals and biotech foods, debilitating the world's population. This could hardly be considered good business except that those who fall ill become customers for partners and subsidiaries in the major drug companies. So, more than any other situation, this is what I really consider threatens our future and makes the present world situation resemble a 'house of cards'. Commercial interests are on the whole geared towards taking and profiting and, like locusts, they contribute nothing *towards renewing the earth*. If the soil fails—they fail. Such renewal is only possible when people work with nature's processes in a harmonious way.

Genetic engineering and nanotechnology

While the relationship between organic and conventional agriculture has not always been an easy one for organic producers, the problem of chemical contamination is a good deal more straightforward than that of biological or genetic contamination. It has been obvious from the outset that GM and non-GM farming cannot exist together without the latter being compromised. In particular, organic farming, which has up to now excluded GM inputs of any kind, will face greater threats in the future unless public resistance to GM foods, in Europe at least, is robustly maintained. The fact that the EC now allows 0.9% (adventitious) GM contamination before recording of GM content on food labels becomes mandatory is simply the thin end of a wedge.

Whatever is said to the contrary, the argument for transgenic crops in agriculture is founded on a business model. Furthermore, the adoption of transgenic technology by governments and research institutes is driven by the need to remain at the forefront of a technology offering commercial benefits and jobs for the host country. GM in agriculture is also substantially underscored by its application in the medical field. There is just so much commercial power behind transgenic technology that, no matter what difficulties and contradictions lie in its path, it bulldozes its way forward. In this respect I see a parallel between the current situation and the absurd genetic science of Lysenko at the time of Stalin, with its ruthless suppression of all dissenting voices. The same tendencies always emerge in human nature whenever power is to be exercised over people—for example the Roman Church altered or suppressed early truths of Christianity in order to consolidate the power of its male-dominated clergy. The modern commercial situation therefore recalls the notable saying about power corrupting.[21]

This brings us to nanotechnology. Here, minute particulate material less than 90 nm (a little over the size of most atoms and about the size of protein molecules) is used to alter the nature of the material into which it is introduced. The mindset, if not the methods of genetic engineering, are brought to mind. On the precautionary principle, organic standards outlaw the use of this technology but in practice widespread contamination by these particles is inevitable. Claims are being made of admirable applications in many fields, antibacterial functions for instance. But the use of nanoparticles to ensure nutrients such as vitamins pass directly into the bloodstream confirms we are treating organisms like machines to be regulated. From a biodynamic standpoint, the use of nanoparticles will only isolate the world of nature from its cosmic origins. At present there is no regulatory framework in place which might guard against unintended outcomes of the deployment of nanoparticulate material. The inhibition of essential microorganisms would, for example, have untold repercussions. Sensitive to the accusation of being a Luddite, one is nevertheless suspicious that this technology is merely waiting in the wings to assail the human race with its own versions of BSE or thalidomide, as well as exacerbating the problems discussed in Chapter 10.

Distortion of agriculture by biofuels

The search for alternatives to the use of fossil fuels is one of the most consuming activities at the present time. But while each of the various alternatives from wind, wave, tidal and nuclear power to photo-voltaics and biofuels generates energy on its own terms, they generate problems as well. The fact that biofuels have an immediate impact on land use demands further comment.

Biofuels claim to be carbon-neutral so are clearly self-defeating the more they depend on chemical inputs. Energy is achieved by different processes. One route is to make biodiesel from vegetable oils. In different parts of the world this is obtained from soybeans, maize, sunflower, canola, rapeseed, palm and jatropha, while waste cooking oil is now being reclaimed for this purpose. Alternatively, ethanol can be made from sugar, starch and, at higher cost, from cellulosic material. For this reason sugar cane, sugar beet and even maize and wheat are candidates.[22] Various forms of short-rotation forestry or coppice systems can also be harnessed for thermal power or heat generation. Leaving aside forestry options, all have in common a consumption of land which might otherwise be used for food crops or animal grazing.[23] At a time when global demand for agricultural commodities has never been greater, and with harvests increasingly threatened by climatic unpredictability, the implications for food scarcity and food prices are clear and have already started to bite. But the issue of competition for land is more complex than this.

There is increasing global demand for meat products and it has long been recognized that in terms of land used animal production is a far less efficient way of generating protein than arable cropping.[24] Expansion of biofuels inevitably impacts on animal husbandry. Less area outdoors signals an increase of intensified (non-organic) production indoors, while less available land for producing feed crops leads to higher costs of animal feeds. So how are these various issues likely to impact on organic farming?

Clearly, there are now greatly increased commercial pressures on land across the globe (see below). Locally, this may increase the tendency to monoculture and pest incidence, while farmers who might have considered organic conversion may now instead choose the energy crop route. One must also hope that the number of organic producers can be maintained in the present economic circumstances, otherwise certified animal feed costs could be the tipping point for many other producers. On the other hand, security of animal feed becomes a powerful incentive for organic farmers to develop truly self-contained farm organisms (Chapter 3).

Major land purchases

Economic history provides a catalogue of human suffering. Within the past century, dam projects and mineral extraction have disrupted the lives of local people, mostly without adequate compensation. In Britain we can go back to the eighteenth century when the Highland clearances and enclosure movement were displacing substantial rural populations. Recently, attention has been drawn to major acquisitions of land in the world's poorer countries.[25] The principal effect of such 'land grabbing' is to concentrate food production ever more in the hands of multinationals and agribusiness. While this situation destroys rural livelihoods and prospects for local food security it will also reduce the future scope for sustainable organic farming systems unless something approaching a new land charter can be devised that secures the rights of local people.

Hybrid farming systems

We have so far spoken of agriculture as conventional or organic for purposes of drawing attention to major differences in approach. However, the organic movement must be well aware that due to costs of production and concerns about the environment, agriculture is not so polarized as it once was.[26] In fact, the last 25 years has seen the emergence of a distinct 'low-input' farming movement, the origins of which lie in 'integrated pest management'.

Clearly we cannot discuss here how such a system works in practice, but because it combines chemicals with ecological methods it is not acceptable in the context of this book. In the author's view it is simply a financial compromise, for the benefits of any ecological methods stand to be cancelled out by the effect of chemicals. Promoted originally by commercial interests in partnership with government agencies, it is basically a business model responding to the increasing cost of chemicals. To be fairer, we can say that it represents a step away from unreformed use of chemicals and may, for the individual or farm in question, constitute a bridge to more ecologically sustainable practices in the future. For the time being, however, it is a home for those lacking true commitment to agro-ecology, or for large-scale situations perceived as too problematic for the introduction of organic methods. Either way, one feels it should be the aim of the organic movement to reach out to such farmers and encourage them to understand what a healthy soil really means and that we must now be prepared to take on certain responsibilities for the sake of the environment.[27]

The trading environment for organics

The market, of course, is the greatest challenge of all. In very simple terms the market for organic products exists only in rich countries or among the wealthy in others. So when wealth diminishes there is an obvious consequence—a consequence felt across the whole world. This underlines the importance of getting away from wealth as an issue governing access to healthy food.

But besides 'certified organic', other trading niches and customer assurances have emerged in recent years as seen on a host of packaged items. In many cases organic or Demeter quality is accompanied by other symbols of customer assurance, making labels often appear overcrowded. The Fair Trade logo has powerful appeal to a public increasingly concerned about exploitation of people in developing countries while the issue of animal welfare has recently received recognition on food products by the RSPCA. Use of the description 'free range', mainly for fresh produce, without necessarily being too clear in meaning, lays claim to a serious market position, while farmers markets and the sale of 'local' produce (mostly not organic) also compete for the purse of those preferring diversified shopping.

While significant reductions in sales of organic produce were being reported at the time of writing, it is worth commenting that the weakening profile of organic lines in supermarkets partly reflects a battle between stores to maintain profitability. While sending a shock-wave through organic marketing, this should be a stimulus for organic producers to realign with independent and local marketing schemes which are undoubtedly their birthright and where the future will be. I also believe that increasing consumer consciousness will be a factor helping *all* involved in ethical trading.

The challenge for biodynamics: reaching a wider constituency

In the foregoing we have identified different kinds of challenge which the organic movement has to face in the years ahead. Meanwhile, biodynamics has lately enjoyed greater publicity and is becoming more recognized in the UK, together with its associated Demeter products. But demand generally outstrips capacity to supply, so for this reason if for no other an expansion of biodynamic farming is needed both in the UK and overseas.

As was mentioned in the first chapter, biodynamics has spread around the world through connection with Rudolf Steiner's social and educational initiatives and to the credit of particular individuals. It is timely for it now to be disseminated more widely within the established organic movement. There are three principal reasons for saying this. In the first place, organic systems depend on life processes for the support of plant growth, these drawing on unseen forces as well as physical substance. In this respect, biodynamics is uniquely equipped to explain aspects of organic farming and to place the farm in the context of processes affecting the whole earth. Secondly, organics on its own may not be enough to meet future challenges such as those of degraded lands or adverse climate. For example, the large size of many farms and the logistics of extensive organic management[28] mean that restoration of soils to a healthy state will require support from combinations of biodynamically inspired preparations. Finally, human life is increasingly at risk from the deadening effects of entirely materialistic thought. As explained in Chapter 1, compared to conventional farming, organics should not merely be a transfer of technology. In reality it is an art. All true art needs the discipline of the intellect but originates from the soul and spiritual aspects of life. The deeper philosophy underlying biodynamics recognizes and supports this essential attribute of the human being.

There are, of course, issues which act as constraints. Many are aware of biodynamics but comprehension of it is not easy—as readers who have survived to this point would concede! Unfortunately, in our age with its dash for quick answers, the essence of biodynamics is all too easily misrepresented or distorted. This recalls the remark by Walter Bagehot: 'One of the greatest pains to human nature is the pain of a new idea.'[29] So expanding the influence of biodynamics is at least partly a matter of effective communication and being able to explain clearly what we understand to exist behind the visible world of nature. It is also about demonstrating that biodynamics works on the ground and is reflected in terms of food quality, nutrition and health. But aside from grasping conceptual issues or gaining awareness of its capabilities, what impediments might there be to the wider adoption of biodynamics?

Practical and cultural issues

To begin with, in BSE-affected areas there are continuing restrictions on access to essential bovine ingredients for the preparations, although these can still be obtained from countries which were unaffected.[30] While some farms do produce their own preparations, access to the necessary raw materials, as well as available time, inhibits on-farm production of both horn and compost preparations and means that practitioners either make them collectively or purchase them from the Biodynamic Association. Overseas, local raw materials for making the horn preparations are widespread, even to having local cattle breeds, but cultural pressures in certain countries inhibit the acquisition and use of animal parts in preparation-making. In this respect, rejection of animistic African and Afro-American practices, the religious beliefs of Hindus and Buddhists and the sentiments of vegetarians are united. The question is: should biodynamics make concessions in these cases?[31] In other areas, animal husbandry can be sparse or non-existent so that few resources are available with which to start up biodynamics. Yet this is no new problem for biodynamics to grapple with, such as its adoption in areas such as Ukraine, where, aside from domestic use, animal husbandry played no traditional part in agriculture.

In addition to the question of animal organs, the requirement to use a common set of compost preparations raises interesting questions. In tropical countries notably, it means importing these unless suitable 'temperate' environments can be found for cultivating the preparation plants. Yet however potent these may be in their areas of origin, they are ill-adapted to the photo-periodism and light intensities of their exotic environments. And because they are usually cultivated rather than grown in a plant community, their vigour is limited. Indigenous plants, even if imperfectly aligned with Steiner's principles, might have a stronger relationship to soil and cosmos than imported ones. And let us not begin to speculate about the effectiveness of imported preparations. So, despite being common practice, neither local cultivation nor importation of finished preparations can be strongly supported.[32] There is no doubt that research into local alternatives should now be encouraged, especially in areas such as India where a strong folk knowledge of plants still prevails relating to healing and planetary relationships. It is vital that all who adopt biodynamics should feel they are 'stakeholders' rather than simply 'followers'.[33]

Aside from any issues relating to compost preparations, it is in relation to the pressures of modern farming life and the truly serious realities of agricultural labour—and people will always compare their own situation with neighbours—that the stirring of field preparations presents a significant deterrent. Although stirring machines and flowforms are available for this purpose,[34] either of these presents a significant cost factor, possibly also at the time of conversion. While they certainly save labour, the *stirring time* still remains an issue. And unless one is to travel with large amounts of stirred preparation, one or other of these devices must be transported. There is no doubt that given good organization and available time, hand-stirring in particular can be a satisfying, social aspect of biodynamics. But this is just one of a number of steps in the making and deployment of preparations, and by no means the only time that a farmer's influence is expressed towards land and crops. From experience in the UK and to a lesser extent overseas, my main concern is that even on the comparatively small scale, farmers are failing to apply preparations in a timely manner or with the frequency that research shows would be beneficial and make the effects of biodynamics more obvious.

A possible solution

In their different ways, the above are matters of preparation supply and application. One should not exaggerate their importance but they are nonetheless deserving of a solution. There might be such a solution which could offer considerable support to the biodynamic movement. In Chapters 5 and 8 the question of potentization was discussed. In the *Agriculture Course* homoeopathy is referred to, though not in relation to the preparations. It was clearly known and recognized by Steiner and was worked on within the biodynamic sphere by his student Lili Kolisko. There is little doubt that since dilution alone is required, potentized preparations (already available) would dramatically cut the time required for deploying biodynamic field sprays. They would also contribute in cases where traditional preparations were unavailable and commend themselves to the treatment of very large areas where only a supply of water is then needed. Although for Demeter certification potentized preparations are not currently permitted as a substitute for the original preparations, the writer considers it timely that this matter is reviewed.

A while back I felt that to open up biodynamics in this way would undermine its principles. But aside from inbuilt conservatism what was the real rationale for this? Homoeopathy is, after all, widely respected among those in the organic movement and many rely on its efficacy to support animal health, if not their own personal needs. Potentized versions of the standard preparations are just that—they start by using exactly the same basic Steiner preparations and are then sprayed in the usual way. Pre-potentized preparations, as tinctures, carry the 'energy' of the substance just as a stirred preparation carries the 'energy' of the original solid preparation (Chapter 5). One can even argue that 'stirring' has already been carried out as part of potentization. Furthermore, a variety of field trials over the past ten years have shown potentized preparations to be effective.[35] Potentized and original preparations do therefore appear to be alternative ways of achieving the same purpose.

Such a measure would, I believe, offer great advantages to a number of existing biodynamic farmers and gardeners while commending biodynamics to a much wider tranche of organic practitioners. Clearly, the intention would not be to displace existing practice but merely provide an allowable option for certified producers. Above all, the spread of biodynamics needs to be facilitated. We need to recognize the world as it is today, and create a better atmosphere for individual research and innovation—that 'personal relationship' to things which Rudolf Steiner sought to emphasize.[36] Ideas either find their time or miss their opportunity.[37]

For this same reason, stronger links should be forged with kindred organizations in the organic and agro-ecological sphere. While biodynamics is small compared with other organizations, much benefit could arise through cooperative working. It is already clear that interest in biodynamics extends far beyond the members of its own Association.

The wider mission of biodynamics

While biodynamics offers a new dimension for the future of agriculture, with further insights it can play a part in tackling social and economic problems as well as informing our attitude towards the future of the earth. Indeed, if human beings seek to engage fully with their own personal development and that of the earth, then biodynamics offers an appropriate nutritional framework.

A new social imperative

At the time of writing, the world is gripped in a financial situation which probably always was the ultimate destination of *laissez-faire* capitalism. For a variety of reasons, the new imperative would appear to be *social responsibility* rather than *individual or corporate opportunism*. It is not difficult to see that rising living costs in combination with reduced disposable incomes will hasten the need for new social and economic forms. This is why in the previous chapter we introduced the subject of community support in agriculture, the ethos of which is substantively based on principles offered by Rudolf Steiner.

As already explained, there are a number of variants, from the more enclosed Camphill model[38] to those where individual farms connect directly in various ways with the surrounding area. Readers will be familiar with sales directly from a farm shop, through farmers' markets or by mail order. Equally well known but often not as local as one might wish are box delivery schemes based on vegetables but including in many cases meat and other consumables. The latter range from standard boxes of different size and content to home deliveries based on a wide choice of available imported produce. While it is difficult to visualize farm delivery models on a scale capable of supporting whole towns or cities, such are already 'up-and-running'.[39] An example is the Teikei system in Japan where consumer groups for each district provide participating farms with a direct distribution network.

Land tenure is a crucial question. Here, a veritable tapestry of social and political traditions confront community farming initiatives. One needs to consider how existing farms are or could be managed. Solutions range from current landownership to the formation of land trusts, or—*in extremis*—more radical transfers of ownership to enable food security for the common good. Meanwhile, the matter of collection, storage, redistribution and delivery is what much of the western world is already good at. In whichever of these situations farms may find themselves there is mutual benefit—gains from financial security as well as from varied types of human contact. So as times change, those who have hitherto dismissed ventures seen as cooperative, collective or communal in emphasis will probably live to see them as a saviour of society. To quote from Helena Norberg-Hodge: 'Promoting links between producers and consumers will not only give us healthier food and a cleaner environment, but will breathe new life into our communities—this is not just about agriculture, it is about the very fabric of our future'.[40]

Reconnection with the land

The human race, now predominantly urban, has largely lost touch with the land, and, indeed, to a considerable degree is without much sense of surrounding community. Many currently yearn for something different from their urban workplace while land work is sometimes seen as a means of healing the psychological damage of modern society. The contribution of the WWOOF organization should not be underestimated in this context. Could community developments in agriculture become a way of bridging the gulf between sections of society who consider themselves separate—urban and rural, disadvantaged and wealthy, youth and the elderly? Might community agriculture and its related activities, such as equipment sharing, provide in future times the crucial foundation for fulfilling work and personal growth? It cannot be emphasized too strongly that society must restore respect for those who work the land and whose toil produces our food. This might well be a positive outcome of a period of high unemployment.

These are questions for society in its search for new ways ahead, but already one can point to important initiatives—different in emphasis but converging on problems of our times. Following Jamie Oliver's trail-blazing school dinners initiative, the 'Food for Life' partnership is now offering schoolchildren the opportunity to grow, cook and eat food from local farms they have visited. No doubt this will sow seeds of a future change of attitude both to food and the land. The Forest Schools programme also recognizes the need for youth to find its place in the natural world as part of becoming a balanced human being.[41] Arising from Rudolf Steiner's understanding of human development is the Ruskin Mill Educational Trust. Here, an open community institution addresses the problems of urban society, providing a holistic training for youngsters with a troubled past. Among its many elements is land work centred on biodynamic horticulture and farming.[42] It is not alone in tackling such questions, for The Prince's Trust has for some 30 years been addressing problems faced by Britain's youth. Increasingly, localization of economic activity is being explored by Local Exchange Trading Systems while under the slogan 'From oil dependence to local resilience' the Transition Town movement is orientating people's minds towards community-building and sharing, rather than emphasis on a consumer culture.[43]

The deeper aspect of our involvement with nature was explored in Chapter 9, where it was found that our 'outer and inner landscapes' are related, and that there are benefits—spiritual benefits—to be gained through reconnecting ourselves with nature and the earth. The truth is that human beings at this time, whether they feel able to articulate it or not, are in desperate need of assurance that there is order and purpose in the universe. Biodynamics, together with kindred paths and the spiritual background upon which they rest, is able to provide such a foundation.

But the idea of drawing closer to nature is not merely for gaining therapeutic advantage or finding pleasant places to go weekend walks. We must realize there is a common bond among all those who care for the earth and humanity. In this respect, those actively committed to organic agriculture and biodynamics are united with the many involved with countless other aspects of nature, conservation and sustainable living—*in this there is no exclusivity*. As individuals or through organizations, human beings in this way give support to the earth's elemental forces. These we have mentioned in earlier chapters as lying behind all nature's outwardly manifested forms. Like the outer reality of nature, this 'elemental world' is needful of our attention at this time and can 'read' our attitudes and ideals. This is the shadow side of the environmental crisis. Were our lack of awareness and concern for this hidden reality to continue, it would be a further cause of our environment becoming daily more hostile. It does not matter that only a relatively small number of people can answer this call—only that such people consciously foster and deepen their understanding.

In conclusion, one can visualize an important synergy arising, for new social and economic developments could enable larger numbers to reconnect with the living earth and her rhythms and, in so doing, contribute to a healthier future.

Appendix: Biodynamics Contacts and Publications

The worldwide biodynamic movement is represented by the Agriculture Section of the School of Spiritual Science at the Goetheanum; Hügelweg 59, CH-4143 Dornach, Switzerland,
email sektion.landwirtschaft@goetheanum.ch.

A selection of UK and overseas contacts are listed below. For countries not listed, go to *links* in **either** the Demeter International website www.demeter.net, email info@demeter.de **or** the UK Biodynamic Association's website below. Biodynamic associations usually have separate addresses from Demeter certification offices.

Australia: Biodynamic Agriculture Australia, PO Box 54, Bellingen, NSW 2454. www.biodynamics.net.au
Email bdoffice@biodynamics.net.au. Also Bio-Dynamic Agricultural Association of Australia, c/o Post Office, Powelltown, Victoria 3797. www.demeter.org.au

Canada: Demeter Canada, 115 Des Myriques, Catevale, Qué JOB 1WO. www.demetercanada.com Email laurier.chabot@sympatico.ca

Czech Republic: Biologisch Dynamisch Sektion, Nemocnicni 53, PO Box 116, CZ 78701 Sumpech. Email pro-bio@pro-bio.cz

Denmark: Demeterforbundet I Danmark, Victor Benixgade 4,2 DK2100, København, *also* Birkum Bygade 20, Odense S, DK 5220. Fax ++45 65 97 30 50

Egypt: Egyptian Biodynamic Association, Heliopolis, El Horrya, PO Box 28 34, Cairo. Email sekem@intouch.com

Finland: Biodynamiska Föreningen, Uudenmaankatu 25A, SF 00120 Helsinki. Fax ++35 89 80 25 91

France: Maison de l'Agriculture Bio-Dynamique, 5 Place de la Gare, 68000 Colmar. www.biodynamie.org

Germany: See Demeter International above.

India: Biodynamic Agriculture Association of India, 31 Signals Vihar, Mhow, MP 453442. Fax ++91 7324 7 31 33

Ireland: Biodynamic Association in Ireland, The Watergarden, Thomastown, Co Kilkenny. Email bdaai@indigo.ie www.demeter.ie

New Zealand: Biodynamic Farming and Gardening Association, PO Box 39045, Wellington Mail Centre. www.biodynamic.org.nz Email biodynamics@clear.net.nz

Norway: Biologisk-Dynamisk Forening, Skippergt 38, N 0154, Oslo 1. Email kontor@debio.no

Slovakia: AJDA Drustvo Za Biologsko-Dinamicno Gospodarjenje, Vrzdenec 60 p, Horjul, SLO 1354. Fax ++386 61 74 07 51

South Africa: Biodynamic Agricultural Association of South Africa, PO Box 115, Paulshof 2056. www.bdaasa.org.za

Spain: La Asociación de Agricultura Bio-dinámica de España. *Secretaria y Tesoreria*, Casa San Martín, Matabuena—40163, Cañicosa (Segovia). biodinamica@terra.es www.assoc-biodinamica.es *Certificación Demeter:* info@demeter.es

Sweden: Biodynamiska Föreningen, Salta, S 15300 Järna, *also* Svenska Demeterförbundet, Skillebyholm, Järna, S 15391. Fax ++46 85515 79 76

UK: Biodynamic Agricultural Association (BDAA), Painswick Inn Project, Gloucester Street, Stroud GL5 1QG. Tel/Fax 01453 759501. Email office@biodynamic.org.uk www.biodynamic.org.uk Demeter certification: Email demeter@biodynamic.org.uk.

USA: Biodynamic Farming and Gardening Association, 25844 Butler Road, Junction City, OR 97448. www.biodynamics.com Email info@biodynamics.com

Biodynamic literature

For supply of ***biodynamic*** books and a range of related literature, contact the Biodynamic Association (UK). See BDAA online shop.

Information and resources on ***organic*** gardening and agriculture can be obtained from *Garden Organic*, Ryton, Coventry CV8 3LG, www.gardenorganic.org.uk, and from *The Soil Association*, Bristol House, 40–56 Victoria Street, Bristol BS1 6BY, www.soilassociation.org.uk

Biodynamic journals

Star and Furrow, Biodynamic Agricultural Association, UK

Lebendige Erde, Germany

Harvests, Biodynamic Farming and Gardening Association of New Zealand

Biodynamics, Biodynamic Farming and Gardening Association, USA

News Leaf, Biodynamic AgriCulture Australia (New South Wales)

Biodynamic Growing, Australian Biodynamic Association (Victoria)

Elementals, Biodynamics, Tasmania

Biodynamis, France

Biodynamic calendars (annual publications for the northern hemisphere)

The Biodynamic Sowing and Planting Calendar, compiled by Maria and Matthias Thun, Floris Books. (Based on GMT)

Gardening and Planting by the Moon, written and compiled by Nick Kollerstrom, Foulsham. (Based on GMT)

Stella Natura: working with cosmic rhythms, Camphill Village Kimberton Hills. (Based on US Eastern Standard Time—'Daylight Saving Time' when appropriate)

Biodynamic training and international forum

In the UK the Biodynamic Association offers a two-year Apprentice Training Programme and a variety of training workshops (www.biodynamic.org.uk). Emerson College runs a Biodynamic-Organic training with winter coursework and summer placements. A short summer course is also held (www.emerson.org.uk).

The USDA website on Education and Training Opportunities in Sustainable Agriculture has listings for other countries. Examples include www.pfeiffercenter.org www.ceres.org.au and www.taruna.ac.nz. For further information and discussion there is the subscription website www.bdnow.org, an international forum on biodynamics for those seeking information or wishing to exchange ideas.

Notes

Chapter 1

1. W.C. Curry, 'The doctor of physic and mediaeval medicine', in *Chaucer and the Mediaeval Sciences*, Allen and Unwin, 2nd edition 1960.
2. An issue raised on many occasions by Steiner. For an accessible source, see Roy Wilkinson, *Rudolf Steiner: An Introduction to his Spiritual World-view*, Temple Lodge, 2001, Ch. 3.
3. R.T. Smith, 'Indigenous agriculture in the Americas', in *Latin American Development*, ed. D.A. Preston, Longman, 1987, pp. 34–69.
4. Many texts include this topic but the work that most influenced the writer was Edward Hyams, *Soil and Civilization*, Thames and Hudson, 1952.
5. W.C. Beets, *Raising and Sustaining Productivity of Smallholder Farming Systems in the Tropics*, Ag Be Publishing, 1989, and D.M. Glen et al., *Ecology and Integrated Farming Systems*, Wiley, 1995. Land equivalent ratio is also well covered on the internet.
6. This would include the fact that mycorrhizal fungi primarily developed from one crop plant have the capability of spreading to benefit others.
7. For the English landscape a primary source is W.G. Hoskins, *The Making of the English Landscape*, Hodder, 1967. For Romania's Transylvanian region, see for example 'The Saxon villages of Transylvania', www.kimwilkie.com.
8. See *Science and Civilisation in China*, ed. Christopher Cullen, Vol. 6, 'Biology and Biological Technology', Part 2 on 'Agriculture' by Francesca Bray, Cambridge University Press, 1975. See also F.H. King, *Farmers of Forty Centuries, or Permanent Agriculture in China, Korea and Japan*, Jonathan Cape, London 1926.
9. Rohan Sisirakumara, Lanka ephemeris and almanac (annual) (lankalit@sltnet.lk). The word *neketh* (Singhala) is derived from *Nakshashtra* (Sanskrit) referring to the days of the moon's sidereal cycle.
10. E. and L. Kolisko, *Agriculture of Tomorrow*, Kolisko Archive Publications, 1978; N. Kollerstrom, *Planting by the Moon*, Prospect Books, 1999, Ch. 3 (refers to the work of Brown and Chow); R. Hauschka, *The Nature of Substance*, Rudolf Steiner Press, 1966.
11. I am grateful to Dr Rohana Ulluwishewa for providing examples of indigenous pest control in Sri Lanka.
12. S. Rajimwale, 'Agnihotra farming method', *Star and Furrow*, 103, pp. 30–1, and P. Tompkins and C. Bird, *Secrets of the Soil*, Viking, 1991, Ch. 19.
13. Steiner, *Nature Spirits*, selected lectures, Rudolf Steiner Press, 1995, and *Harmony of the Creative Word*, Rudolf Steiner Press, 2001. Also M. Pogacnik, *Nature Spirits and Elemental Beings*, Findhorn Press, and *Nature Spirits and What They Say*, interviews with Verena Stael von Holstein, edited by W. Weirauch, Floris Books.
14. A classic of its kind was Jacks and White, *The Rape of the Earth*, Faber, 1939. See also M. Perelman, 'The Green Revolution', in *Radical Agriculture*, Harper and

Row, 1976. A useful section on the Green Revolution is to be found in *The Seed Savers Handbook*, Grover Books, 1996, Ch. 2. A review of the changes towards intensification in the UK is provided in P. Conford, *The Origins of the Organic Movement*, Floris Books, 2001.

15. F. Chaboussou, 'How pesticides increase pests', *The Ecologist*, 16 (6), 1986, 29–34. The designation C3 or C4 refers to the biochemical pathway by which sunlight is converted into simple sugars and broadly equates with the light intensity of temperate and tropical conditions.
16. For a full listing of permitted pesticides, see *Pesticides Register* on the internet.
17. For the historical development of organic farming one cannot do better than refer to Philip Conford's *The Origins of the Organic Movement*, Floris Books, 2001, and his earlier anthology *The Organic Tradition*, Green Books, 1988. A stimulating discussion is also to be found in *Organic Futures* by Adrian Myers, Green Books, 2005.
18. *Permaculture* was originated in the 1970s by the Australians Bill Mollison and David Holmgren. See Wikipedia online.
19. Organic farming organizations have all contributed to the literature on this subject. In the USA the Rodale Institute is notable. Nicolas Lampkin's *Organic Farming* (1990) has stood the test of time while Herbert Koepf's *The Biodynamic Farm* and Sattler and Wistinghausen's *Biodynamic Farming Practice* (Biodynamic Agriculture Association, 1992) should be referred to.
20. Companion planting is a traditional part of organic gardening and evokes some of the ancient systems. We need especially to consider it in the context of biodiversity and rotations (Chapter 3).
21. Arising out of long experience with biodynamic gardening and natural control methods, John and Helen Philbrick produced a classic work, *The Bug Book* (3rd edition 1963). This and much other related information is available via the internet.
22. For an accessible source of information on the many substances contained in plants, see the Soil Association's *Organic Farming, Food Quality and Human Health—A Review of the Evidence*, 2002.
23. Patrick Holden encapsulated this idea when recently he said that biodynamics represented the 'spiritual conscience' of the organic movement.
24. The term first used in the German language was Biologisch-Dynamisch Landwirtschaft.
25. Rudolf Steiner, *Agriculture Course*, Rudolf Steiner Press, 2004.
26. Eve Balfour, *The Living Soil and the Haughley Experiment*, Faber and Faber, 1975 (originally published as *The Living Soil,* 1944). See also Albert Howard, *An Agricultural Testament*, Oxford University Press, 1940.
27. See note 13.
28. Worldwide there are some 4300 farms registered *Demeter*, comprising around 130,000 ha. See *Demeter International Statistics*, 2008. For information on the range of activities based on the work of Rudolf Steiner in the UK, see the booklet *Anthroposophical Research and Endeavour,* available from rsh-office@anth.org.uk.
29. See for example C. Bigler et al., 'Rotwein unter Hochspannung: Mehrjährige

Qualitäts—Untersuchung mit Gas-Discharge-Visualisation (GDV)', *Beiträge zur 10 Wissenschaftstagung Ökologischer Landbau*, Zurich 11–13 Feb. 2009, Band 2: 441–3. www.orgprints.org/14446/. (A comparative study of red wine quality.) See also Markus van der Meer et al., *Lebendige Erde* (Sept. 2009).

30. Alex Podolinsky was the pioneer of biodynamics in Australia and published, in two volumes, *Biodynamic Agriculture Introductory Lectures*, Gavener Publishing, Sydney, 1985/89. Peter Proctor has likewise elevated biodynamics in New Zealand, leading to a series of recent publications: *Biodynamics: New Directions,* 1989; *Grasp the Nettle,* 1997; and *Biodynamic Perspectives*, ed. Gita Henderson, 2001. See *Star and Furrow*, 107 (2007), 18–21 for a feature on Peter Proctor's initiative for biodynamics in India. See also the DVD entitled *One Man, One Cow, One Planet.*
31. Ibrahim Abouleish, *Sekem*, Floris Books, 2004.
32. Pioneering work has been conducted by Daycha Siripatra in Thailand and Nikanor Perlas in the Philippines (see internet). Progress has also been made in Sri Lanka as shown by the recent increase in numbers of farms individually or group registered with Demeter. See R.T. Smith, *Star and Furrow*, 203 (2005), 34–6.

Chapter 2

1. This is to do with the process of incarnation and excarnation, which governs our present stage of evolution on the earth. See Steiner, *An Outline of Esoteric Science*, Anthroposophic Press, 1997.
2. Roy Wilkinson, *Rudolf Steiner: An Introduction to his Spiritual World-view.*
3. Sherry Wildfeuer, *What is Biodynamics?* See www.steinercollege.org/anthrop/sn.html.
4. The current market for atheistic books continues unabated. See, for example, Richard Dawkins, *The God Delusion.* Paradoxically his anger at the religious concept of God palpably exposes the frustration of not being able to solve this problem in a scientific way.
5. A wonderful biography of C.G. Jung is that by Laurens van der Post, *Jung and the Story of Our Time*, Hogarth, 1976.
6. This has been termed 'logical positivism'. Intellectual development, originating from Aristotle, finally separates from any spiritual foundation in Kant's *Critique of Pure Reason.* Refer to A.J. Ayer, *The Problem of Knowledge.*
7. Note 2.
8. See, for example, Steiner, *Nature Spirits*, Lecture 1, p. 21.
9. Karl König 'Embryology and world evolution', *British Homoeopathic Journal* (reprint of series of articles, 1968–9).
10. A.R. Wallace, *Man's Place in the Universe*, 1903. Such ideas on evolution were carried into the social sphere by Herbert Spencer (see Chapter 11). For discussion of the problems of evolutionary theory, see articles by Paul Carline and Terry Goodfellow in *New View*, Issue 51, June 2009.
11. M. Rossignol et al., 'Lunar cycle and nuclear DNA variations in potato callus', in *Geocosmic Relations*, eds Tomassen and Pudoc, Netherlands, 1990, pp. 116–26.
12. *Popol Vuh—The Great Mythological Book of the Ancient Maya*, ed. R. Nelson,

Houghton Mifflin, 1977.

13. Gautama, the Lord Buddha, expressed these ideas very explicitly.
14. This being was also recognized in antiquity as Ahura Mazda (by the ancient Persians), Apollo and Dionysus (in Classical Greek times). For further information, see Steiner, 'The being of Anthroposophia', lecture, Berlin, 3 February 1913.
15. Various translations of the *Bhagavadgita* have appeared over the last century. The one I use here is a translation by S. Prabhavananda and C. Isherwood. It has an introduction by Aldous Huxley which is worth reading on its own account.
16. This is despite what we might think of our modern world! But with increased freedom and access to material wealth comes responsibility—moral responsibility.
17. For insight into Druidism, see P. Carr-Gomm, *Druid Mysteries*, Rider, 2002.
18. For an understanding of how the Greeks viewed these conditions, see Steiner's *Warmth Course*.
19. Kepler had experience of this, for—as reported in Norman Davidson's *Sky Phenomena*—in relation to planetary movements, he said, 'I climb along the harmonic scale of the celestial movements to higher things where the true archetype of the fabric of the world is kept hidden.'
20. Steiner tackles the history of the earth in *Outline of Esoteric Science* and *Cosmic Memory*. He makes clear that the evolution of the Earth and of humanity are one and the same, and that we have actually 'descended' from a spiritual or supersensible state rather than 'ascended' from physical substance. His account of the roles of different spiritual hierarchies in the forming of existence is briefly set down in *Genesis*, Rudolf Steiner Press, 2002, pp. 76–9.
21. 'Chaos and cosmos' is covered in *Agriculture*, extracts from Rudolf Steiner, edited by R.T. Smith, Sophia Books, 2003, pp. 123–30 and References. In Diana Cooper and Shaaron Hutton's 'angel-inspired' *Discover Atlantis* we read that originally the 'Source' sent out images of itself to all parts of the universe to experience and grow, and to bring back their experiences to the Godhead. Such an undertaking has required an 'Intergalactic Council made up of cosmic energies' (note 31). A number of ideas are in accord with Steiner and Theosophical literature, though the language used is quite different and the meanings are often vague.
22. Laurens van der Post wrote a number of travel books, several about Africa, the most relevant here being *The Lost World of the Kalahari*, Penguin. The event referred to was featured in a famous documentary film made in the 1970s.
23. I am grateful to Adriana Koulias for an article in the *Anthroposophical Newsletter*, Oct. 2007, Vol. 84, No. 5, pp. 4–5, which explains that such divine female forces have worked throughout human history and have a connection with the constellation of Virgo. We will note in Chapter 6 that this constellation is connected with the *earth* element.
24. J. Lovelock, *Gaia—A New Look at Life on Earth*, Oxford University Press, 2000.
25. Various Steiner sources, but I suggest here *Spiritual Beings in the Heavenly Bodies* and *Mystery of the Universe*. See also Wolf Storl, *Culture and Horticulture*, Ch. 4, and, more basic, M.A. Sanders, *Astrology*, Park Lane Press.
26. For further discussion of the atom, see Steiner, *The Temple Legend*, Rudolf

Steiner Press. See also *Agriculture*, extracts from Steiner compiled by R.T. Smith, Ch. 5, p. 96.

27. Steiner, *Mystery of the Universe*, Rudolf Steiner Press, pp. 58–9.
28. See Olive Whicher, *Sunspace,* and Lawrence Edwards, *Projective Geometry*. The latter may be defined as a branch of mathematics which takes as its basis the three primary elements of point, line and plane. See also Nick Thomas, *Space and Counterspace*, Floris Books, 2008.
29. Reiki and other forms of healing use crystals and there is widespread belief in their power.
30. Here, I am grateful to Howard Smith for a point of clarification on Steiner's *Warmth Course*. See Rudolf Steiner, *Science, An Introductory Reader*, extracts compiled by Howard Smith, Sophia Books, Ch. 10 and endnote 36.
31. Akashic records are a spiritual history of all past events. Angel-inspired writings include Diana Cooper and Shaaron Hutton, *Discover Atlantis*, and Diana Cooper, *Angel Inspiration*, Hodder.
32. Rupert Sheldrake is known for his pioneering approach to what he termed 'morphic fields' or 'morphic resonance'. *Dogs That Know Their Owners are Coming Home* is one of his better-known books.
33. Lynne McTaggart, *The Field*, Element Books, 2001. Reference to such a concept is also found in *The Mystery of the Crystal Skulls* by C. Morton and C.L. Thomas, Thorsons, 1997.
34. Steiner, *The Influences of Lucifer and Ahriman*, Lecture 5, SteinerBooks, 1993, p. 80.
35. John Gribbin, *The Universe—A Biography*, Penguin-Allen Lane, 2006.
36. Temperatures in physics are normally referred to in the Kelvin scale which starts at absolute zero, or $-273°C$, the temperature below which matter cannot be taken.
37. See NASA website and, for an example, www.ucar.edu/communications/highlights/1996/research.html.
38. Steiner, *Mystery of the Universe*, Lecture 4, pp. 48–9.
39. *Stella Natura*, 2001 (Kimberton Hills biodynamic calendar), pp. 10–11, article by Robert McCracken entitled 'The star we live by'.
40. For the origin of some of these ideas, see *The Living Universe and the New Millennium*, lectures by Willi Sucher, Anastasi. This process of energy accumulation and light production in the solar corona mirrors what has been proposed to account for quasars (quasi-stellar sources) which lie in front of black holes.
41. NASA website. See also *Results from the Biodynamic Sowing and Planting Calendar* by Maria Thun, Floris, 2003. The experience of bakers is instructive on day-to-day variations and may occasionally reveal the consequences on yeasts of lunar perigee, nodes and eclipses.
42. Steiner, *Agriculture Course*, Lecture 6.
43. Norman Davidson, *Sky Phenomena* (Floris Books, 1993), is easy reading and highly recommended.
44. Miles Mathis, 'Explaining the ellipse', www.wbabin.net/mathis/ellipse
45. Willi Sucher, op. cit. (note 40) and Brian Keats, *Betwixt Heaven and Earth*, 1999, pp. 120–30.
46. Steiner, *Mystery of the Universe*, p. 56.

47. Steiner, *Inner Impulses of Evolution. The Mexican Mysteries. The Knights Templar.* Seven lectures (GA171), Anthroposophic Press.
48. Bode's law is discussed in Davidson (Note 43) and also Brian Keats, *Betwixt Heaven and Earth*. Rudolf Steiner, in *The Festivals and their Meaning*, points out that the Magi (the Three Kings) were initiates who could read mathematical signs and were guided on this basis.
49. Davidson, op cit. (note 43). The references made to music in this chapter, or more simply to sound, may help us approach a real meaning for the cosmic *Word*.
50. John Michell, *City of Revelation*, Garnstone Press, 1972.
51. Gordon Strachan, in *Prophets of Nature: green spirituality in romantic poetry and painting* (Floris Books, 2008), notes that this 3:4:5 triangle also connects with the ratios of musical intervals. See also his *Jesus the Master Builder*, Floris Books, 1998.
52. See, for example, *Moon Rhythms in Nature* by Klaus-Peter Endres and Wolfgang Schad, Floris Books, 2002.
53. Ernst M. Kranich, *Planetary Influences Upon Plants*, Biodynamic Literature, 1984, pp. 55–6.
54. Gerbert Grohmann, *The Plant*, Vol. 2, Bio-Dynamic Farming and Gardening Association, Kimberton, PA, 1989, pp. 35–6.
55. This is sometimes referred to as the Golden Mean or the Golden Section. It is a proportion which appears throughout the natural world and which the human being appears to recognize as aesthetically pleasing and balanced. Information is available on the Web. My personal reference has been the venerable work by Jay Hambidge, *The Elements of Dynamic Symmetry* (1926), reprinted by Dover Books.
56. In other words displaying the Fibonacci principle. For illustration, see Grohmann op. cit. (note 54) and Keats op. cit. (note 45).
57. Lawrence Edwards, *The Vortex of Life: Nature's Patterns in Space and Time*, 2nd ed. Floris Books, 2006, Ch. 3. See also the excellent website www.nct.anth.org.uk/path.htm or go to Google and type 'path curve'.
58. Margaret Watts Hughes, *Voice Figures*, Hazell, Watson and Viney, London 1891. Patterns were achieved in sand, *Lycopodium* spores, plaster of Paris, water and other media by placing these materials against a resonant elastic membrane to which vocal sounds were directed through a special funnel. More recently, water has been subjected to vibrations of different frequencies, the resulting surface patterns likewise resembling a variety of organic forms. See Alexander Lauterwasser, *Wasser Klang Bilder*, AT Verlag, Aarau-München, Germany, 2003.
59. I am grateful to Piyatissa Wanigasekera for this information. It will be noted by those using the Thun biodynamic calendar that reference is made there to *all* the planets of the solar system.
60. Bernard Lievegoed, *Study Material for Experimental Circle of Anthroposophical Farmers and Gardeners* (originally offered in 1951). See Biodynamic Association. Lievegoed's *Man on the Threshold*, Ch. 17, also offers insight into the role of the planetary spheres.
61. Steiner, *Agriculture Course*. The polarity between calcium and silica is a continuing theme throughout this course. See especially Lecture 3.
62. Bases such as potassium and magnesium, which are highly soluble in water,

contribute to the osmotic properties of cells.

63. Note here for example, Aura Soma.
64. Steiner (see note 61). In the popular work *The Mystery of the Crystal Skulls* by C. Morton and C.L. Thomas (Thorsons, 1997) 'knowledge' is attributed to one or more of the skulls, and it relates that 'silicon was introduced into our genetic structure' and 'is now in our blood . . . within us is a part of the whole crystalline matrix that links us to the rest of the galaxy'.
65. I am grateful to Alan Brockman for sharing this thought with me.

Chapter 3

1. I am grateful to the environmentalist Jonathan Porritt for having made this subtle distinction.
2. Such a view was possible until comparatively recent times, for Johannes Kepler (1571–1630) the astronomer considered the earth to have a soul. See L. Günther, *Kepler und die Theologie*, Giessen.
3. Steiner, *Cosmic Memory*, SteinerBooks. For reference to the earth's rocks having been alive, see also Walter Cloos, *The Living Earth*, Lanthorn Press.
4. Steiner, *The Cycle of the Year*, Anthroposophic Press, 1984. See also Steiner's *Universe, Earth and Man* and the *Agriculture Course*, Lecture 2. In Greek mythology, it was Demeter's daughter Persephone (or Proserpina) who disappears to Hades in the underworld for a season, thus symbolizing a process of withdrawal and emergence.
5. Steiner, *Warmth Course*, lectures, Stuttgart 1920, Mercury Press, 1988.
6. Steiner, *Astronomy Course*, Mercury Press, 1984. The etheric or life quality of the sun is 'suctional', a force which therefore increases in the daytime and in summer. This accounts for the earth's out-breathing.
7. Steiner, *At the Gates of Spiritual Science*, Rudolf Steiner Press.
8. The earth's 'atmosphere' consists of many layers, gaseous below and ionic in its most distant regions some 500–50,000 km above the earth's surface. The lowest layer or troposphere—to 12 km—is the layer in which weather occurs, being the most dense and containing water vapour. Differences in heating, influenced by latitude and distribution of land and sea, lead to variations in atmospheric pressure. These lead in turn to winds, which are subject to the drag effect of the earth's rotation—the Coriolis force—which causes them to turn in opposite directions north or south of the Equator. As we move away from the earth's surface, atmospheric layers become more tenuous and eventually pulsate due to magnetic distortions imposed by the solar wind. See Chapters 7 and 8 in *Gaia Sophia* by Kees Zoetemann. See also Wolf Storl, *Culture and Horticulture*, p. 235.
9. One documented case is that of Edgar Mitchell as reported in *The Field* by Lynne McTaggart, Element Books, 2001. See also Friedrich Benesch, *Ascension*, Floris Books, 1979.
10. Steiner, *Lucifer and Ahriman*, Anthroposophic Press, pp. 74–6.
11. Here, one can choose many sources connecting with organic farming or permaculture. See Nicolas Lampkin's *Organic Farming*, Farming Press. See also Adrian Myers, *Organic Futures*, Green Books.
12. Steiner, *Agriculture Course*, Lecture 2. Two very useful articles on this theme, by

Manfred Klett and by Tadeu Caldas, appeared in *Star and Furrow*, 101 (2004).

13. Steiner, *The Spirit in the Realm of Plants*, Mercury Press, 1978.
14. This may be more a fundamental need, for the ancestor of cattle was the aurochs—a woodland animal—and hence Steiner's remark about cattle having an affinity with woody, broadleaved species.
15. Steiner, *Agriculture Course*, Lecture 2. Clearly, a neighbouring farm's animal dung or feed grains would meet a farm's needs better than materials imported from far away.
16. This cosmic factor is sometimes referred to as astrality. It simplifies the idea if we say that in order for cosmic wisdom or messages (discussed in Chapter 2) to take effect there must be a sensitivity to them within the soil. This sensitivity, or ability to listen to the cosmos, is a property of nitrogen, especially that originating from animal manures.
17. Karl König, *Earth and Man*, Biodynamic Literature, pp. 193–259. Steiner, *Agriculture Course*, Lecture 2. See Chapters 2 and 10.
18. Steiner, *Agriculture Course*, Lecture 4, and *Harmony of the Creative Word*, pp. 10, 25–6. An experiment was reported some years ago from Germany where bull and cow dung had been compared in the growth of vegetable crops. Similar in nutrient content, nevertheless the cow-fertilized plot offered more sustained growth. An observation from Sri Lanka on the herb gotu kola, the pennywort (*Centella asiatica*) grown with biogas digested slurry or cow manure, indicated lush growth with the former but superior flavour resulting from cow manure.
19. Manfred Klett, *Principles of the Biodynamic Spray and Compost Preparations*, Floris Books, 2006, Ch. 2.
20. *Agriculture Course*, Lecture 4. In Sri Lanka, the tethering of a cow to a palm tree for 14 nights was recognized to give the same fertility benefit as tethering a buffalo for 10 nights! This must say something about the tuning of an animal to its native environment. The subject of horns and hooves is discussed in depth by Enzo Nastati. Further Reading (Meeting 14).
21. One can feel sadness when looking at polled cattle where the skull forms a bony bulge unable to form horns. The author has heard several reports of polled cattle attacking horned cattle without provocation or behaving in a threatening manner towards people. This is circumstantial evidence of an imbalance in the cattle organism.
22. For those who take issue with the animal parts used in the compost preparations, it is worth noting that Maye Bruce, a vegetarian from the US, used honey to combine the various herbal ingredients. Her preparation is available as 'Quick Return' (used as a compost starter) from the *Organic Gardening Catalogue*. Regarding certification, Demeter does not approve this as a substitute for the original compost preparations.
23. *Animals and their Destiny* by Karl König. See also Fred Kirschmann, 'Animals in an organic system: exploring the ecological, social and economic functions of animals in organic agriculture, *Ecology and Farming*, Sept.–Dec. 2006, 26–32, IFOAM. See also Nikolai Fuchs, 'The significance of animals for the earth', *Star and Furrow*, 98 (2003), 12–14.
24. Maria Thun, *Gardening for Life*, Hawthorn Press, 1999, pp. 58–65. Also Thun,

Results from the Sowing and Planting Calendar, pp. 136–9.

25. *Agriculture Course*, Lectures 4, 5.
26. *Agriculture Course*, Lecture 7.
27. Here we refer especially to patented organisms such as EM (effective micro-organisms). Without specific need being identified, and particularly in hot countries where rates of composting are rapid, their use is needless.
28. The use of imported organisms is against the spirit of the farm organism and might risk introducing GM material. Steiner considered it misguided to be culturing organisms for this purpose. *Agriculture Course*, Lecture 4.
29. In extreme circumstances where highly unbalanced, toxic or putrefying materials need to be degraded without delay, specially cultured organisms may be applied and appropriate public health precautions taken, but this is hardly organic farming!
30. New Zealand farmers have made good use of various materials. See, for example, *Biodynamic Perspectives*, NZ Biodynamic Association, pp. 38–48. More detail is provided in the author's book produced for Sri Lankan readers, *Organic Farming: Sustaining Earth and People*, Colombo 2007 (available from the author via email). See also Storl, op. cit., Ch. 14.
31. *Compost Tea Brewing Manual* (5th edition) by Elaine Ingham, www.soilfoodweb.com. See also *Teaming with Microbes—A Gardeners Guide to the Soil Food Web*, by Jeff Lowenfels and Wayne Lewis, Timber Press, 2006.
32. Steiner, *Agriculture Course*, Lecture 8. Long before this course was offered, Steiner made the remarks quoted here. See *Agriculture Course*, American edition, Appendix B, p. 250.
33. This view of nature is embodied in Lecture 7 of the *Agriculture Course*.
34. Callum Coats, *Living Energies*, Gateway Books, 1996. A book devoted to the legacy of Schauberger.
35. Robert Graves, *The White Goddess*, Faber.
36. For geodetic phenomena, see *The Pattern of the Past* by Guy Underwood, *The View Over Atlantis*, by John Michell, and *The Ley Hunters Companion*, by Paul Devereux and Ian Thomson. Reference to ultrasound research is found in *Druid Mysteries*, by Philip Carr-Gomm, Rider, 2002.
37. Here, the reader should recall the introductory section of this chapter. See also Steiner, *Cosmic Workings in Earth and Man*, Rudolf Steiner Press, pp. 65–9. To build an understanding of the notion of *astral forces* one should consider mother's milk, that complete food for the development of the young and a product of those intensified forces found within the animal and human organism. Plants do not have these forces internalized, but away from the ground 'astrality' intensifies. An example to illustrate this is the coconut palm where in the heights of the trees the nuts contain a milky fluid.
38. It may be of interest to note that birdlife in the tropics often displays much brighter plumage compared to temperate regions. This could be connected to the strength of astral forces in these regions. A complementary view of insects and trees is offered by Enzo Nastati, *Commentary on Rudolf Steiner's Agricultural Course*, Eureka, Trieste 2009.
39. Steiner, *Bees* (originally *Nine Lectures on Bees*), Anthroposophic Press, 1923,

Lecture 1, p. 20.
40. Steiner, *Agriculture Course*, Lecture 7.
41. The booklet by Martin Pfeiffer, *The Agricultural Individuality—A Picture of the Human Being*, is helpful on the bee and other aspects of this chapter. The work by Günther Hauk, *Towards Saving the Honey Bee* (Biodynamic Farming and Gardening Association, USA 2002), is an informative reference. See also Michael Weiler, *Bees and Honey: From Flower to Jar*, Floris Books.
42. Matthew Barton, 'What is wrong with the bees?', *New View*, 49 (Autumn 2008), pp. 3–6.
43. Steiner, *Harmony of the Creative Word*, Lecture 4.
44. Steiner, *Agriculture Course*, Lecture 6.
45. Perhaps there is a parallel here with our modern commercially driven approach to domestic hygiene. May we not weaken our immune systems by overprotecting them from everyday germs, thereby predisposing ourselves to more serious pathogens?
46. In Chapter 4 we refer to hydrogen as the carrier of the cosmic spiritual force.
47. Steiner, *Agriculture Course*, Lecture 3.
48. Steiner, *Harmony of the Creative Word*, Lecture 3.
49. See Steiner, *Nature Spirits*, and *Harmony of the Creative Word* (both Rudolf Steiner Press). See also *World Ether, Elemental Beings, Kingdoms of Nature*, extracts from Rudolf Steiner compiled by Ernst Hagemann, Mercury Press, NY 1993.

Chapter 4

1. For simplicity's sake, systematic treatment of the ethers was considered to lie outside the scope of this book and a simplified approach is therefore adopted in this chapter. I find helpful the approach adopted in the *Handbook* of the Biodynamic Education Centre, compiled by Lynette West, Level 3, pp. 34–46. See 'The ethers' by Anna Irwin, *Star and Furrow*, 101 (2004), 38–41. Also Olive Whicher, 'Creative qualities in the forces of the sun', *Star and Furrow*, 83 (1994) 29–32. Readers wishing to seriously pursue the subject of ethers should consult the work by Gunther Wachsmuth, *Etheric Formative Forces in Cosmos, Earth and Man*, Anthroposophic Press, 1934, and Ernst Marti, *The Four Ethers*, Schaumberg Publications, 1984.
2. The most important pigment is chlorophyll 'a'. Besides this there are chlorophylls 'b' and 'c', together with carotenoids, which can be observed in autumn colours. Chlorophyll is confined to the photosynthetic or thylakoid membrane of the chloroplast.
3. *Mineral Nutrition of Higher Plants* by Horst Marschner, Academic Press.
4. Steiner, *Bees*.
5. I am grateful to the late Olive Whicher for discussion on etheric spaces. Olive worked with George Adams. See his *Physical and Etheric Spaces*, Rudolf Steiner Press, 1965. See also Lawrence Edwards, *The Vortex of Life*, Floris Books, 2nd ed. 2006, p. 158. The concept of chaos as developed by Steiner is covered in *Agriculture* (extracts from Steiner) Chapter 6, ed. R.T. Smith, Sophia Books. It is also referred to in Chapter 7 herein.
6. 'New insights into cellulose structure', H. Chanzi et al., 2005, www.ill.fr/AR-99/

page/12polym.htm.

7. Jochen Bockemühl, *In Partnership with Nature*, 1981; *Extraordinary Plant Qualities for Biodynamics* (with K. Jarvinen), 2006.
8. Steiner, *Evolution of Consciousness*, Lecture 7, Rudolf Steiner Press.
9. Ernst Lehrs, discussing Goethe in *Man or Matter*. Additionally, we observe that the word *inspiration* means 'in-breathing', an allusion to drawing in ideas from a world of universal knowledge.
10. Various Steiner sources could be quoted. See for example *Genesis*, pp. 70–1.
11. *World Ether*, a compilation by E. Hagemann of Steiner's writings on elemental beings, p. 14.
12. Steiner, *Nature Spirits*, and *Harmony of the Creative Word*. See also *World Ether, Elemental Beings, Kingdoms of Nature*, extracts from Steiner compiled by Ernst Hagemann.
13. Steiner, *Harmony of the Creative Word*, Lecture 7.
14. *The Tempest* is teaming with allusions as the various characters, some with obvious elemental names, express their sentiments. In Ariel's song (that of a sylph) we find the well-known 'Where the bee sucks there suck I—in a cowslip's bell I lie . . .'
15. *Agriculture Course*, Lecture 2.
16. *Agriculture Course*, Lecture 2.
17. *Agriculture Course*, Lecture 2.
18. Steiner, *Harmony of the Creative Word*, Lecture 7, p. 111.
19. Steiner, *Man and the World of Stars*, pp. 118–19.
20. E. and L. Kolisko, *Agriculture of Tomorrow*, Chapters 3 and 4.
21. Steiner, *Mammoths and Mediums*, lectures to workmen at the Goetheanum, Rudolf Steiner Press, p. 212.
22. *Agriculture Course*, Lecture 2 (American translation). For further ideas, see *Silica, Calcium and Clay* by F. Benesch and K. Wilde, Schaumberg Publications, 1995. For an overview of soil properties, see the author's article 'Knowing more about soil', *Star and Furrow*, 105 (2006), 33–8.
23. R. Hauschka, *The Nature of Substance*, Ch. 20.
24. *Agriculture Course*, Lecture 4 (transl. George Adams). Reference to 'those structures in the earth kingdom' in this passage is interpreted here to mean the clay-humus complex, or colloids within the soil.
25. This emphasizes the point already made about the participation of humus as well as clay in holding cosmic energies. Humus in practice is the product of recombination of many breakdown products of plant and animal origin.
26. This comes from a remarkable book based on communication with elemental beings, *Nature Spirits and What They Say*, edited by Wolfgang Weirauch, Floris Books.
27. After hydrogen and helium, carbon, oxygen and nitrogen also happen to be the commonest elements in the sun.
28. While carbon supplies the needs of plants, and of soft tissues, animal life calls upon the support of lime. For Steiner's main comments about the common elements of living organisms, see Lecture 3 of the *Agriculture Course*. See also Enzo Nastati, *Commentary on Dr Rudolf Steiner's Agricultural Course*.
29. R. Hauschka, *The Nature of Substance*.

30. Substances such as copper, zinc, selenium and cobalt occur in low concentration in soil but nevertheless have proven functions in enzyme systems and are therefore termed 'essential micronutrients'. Molybdenum and cadmium play important roles in activating enzymes that govern nitrogen fixation in legume rhizobia. Chromium is not normally regarded as a plant nutrient. At higher concentrations as in soils derived from serpentine rocks it is inhibitory to plant growth, while at low concentrations it can be stimulatory. Chromium is connected with the 'glucose tolerance factor' and is therefore essential in carbohydrate metabolism. S. Samantarai and P. Das, *Acta Physiol. Plantarum*, 20 (2) 1998, 201–12.
31. Ultimately our understanding of the role of elements must acknowledge their contribution to an organism's physical, etheric or astral functioning, and in this respect the vortical presentation of the elements by Glen Atkinson offers a promising hypothesis. See www.garudabd.org. *The gyroscopic periodic table* is accessed *via* the 'books' link. These aspects are already taken into consideration in anthroposophical medicine. Consult, for example, *Anthroposophical Medicine* by Victor Bott, Rudolf Steiner Press.
32. Steiner, *Agriculture Course*, Lecture 5.
33. Transmutation is most likely to involve the creation of a required element using pre-existing ones as raw materials. Steiner also referred to substances existing in the atmosphere and outside the earth in homoeopathic dilution.
34. R. Hauschka, *The Nature of Substance*, Ch. 3, and also his *Nutrition*, p. 12 (both Rudolf Steiner Press). Herzeele's key work was 'The Origin of Inorganic Substances' (in German, Berlin, 1876).
35. Louis Kervran, *Biological Transmutations*, Crosby Lockwood, 1971.
36. J.W. and E.A. Moore, *Environmental Chemistry*, p. 5. Texts on geochemistry are useful for such data, one source being Rankama and Sahama, *Geochemistry*.
37. We are concerned here with atomic mass—the size of the nucleus—for this is what defines the 'individuality' of each element. Electronic configuration, determining behaviour in chemical reactions, is due to polarization of the atom resulting from electromagnetic and gravitational laws of the mineral world.
38. A useful general source on Steiner's conception of earth history is his *Cosmic Memory*, SteinerBooks. The author's best estimate is that the *Lemurian* age includes the Palaeozoic and early Mesozoic geological periods.
39. *Agriculture Course*, Lecture 5. This supports the notion that the 'biography' of an element plays a significant part in its value for the health of living organisms. This is a principle essential to the nutritive value of organic and biodynamic agricultural products (see later comments).
40. Failing any other insight, I am inclined to view this form of nitrogen as the relatively short-lived isotope ^{15}N which breaks down to nitrogen-14, releasing a neutron in the process.
41. Referred to by R. Hauschka, *Nutrition*, Ch. 19, pp. 133–4.
42. For example, M. Werner and T. Stöckli, *Life from Light*, Clairview Books.
43. On this subject F. Benesch and K. Wilde in *Silica, Calcium and Clay*, pp. 73f., write more fully, indicating that calcium's role has caused it to lose a certain connection to cosmic influences. This implies that calcium 'fell' from higher

cosmic substance in order to carry out its mission (see equation 2).

44. One might draw a parallel here with that suctional force we encountered when discussing the sun in Chapter 2.
45. Besides the content of Chapter 6, particular insight may be obtained from the discussion of the vortex in Chapter 8.
46. Steiner, *Harmony of the Creative Word.*
47. R. Hauschka, *The Nature of Substance.*
48. Steiner, *Agriculture Course.*
49. R. Hauschka, *Nutrition*, Ch. 24, p. 164. See Chapter 8 herein.

Chapter 5

1. The spray preparations are referred to in Lecture 4 and the compost preparations in Lecture 5 of Steiner's *Agriculture Course.*
2. The significance of the numbers used as abbreviations for the preparations appears to originate from experimentation by Steiner as reported to G. Wachsmuth. We appear to be dealing with a fifth series of trials, as originally documented in the Weleda laboratory.
3. Information is available in many biodynamic sources. Two excellent booklets have been produced by a German team: *The Biodynamic Spray and Compost Preparations 1. Production Methods,* and *2. Directions for Use.* These are obtainable from the BDAA from whom a series of information leaflets is also available.
4. A note on low latitude locations. The general guidance is to select the low sun period for 500 and high sun for 501. Distinctively, these periods may only amount to 3–4 months. It is doubtful whether this has such an effective influence as in temperate regions and in the author's view the key process becomes the diurnal rhythm.
5. F. Sattler and E. von Wistinghausen, *Biodynamic Farming Practice*, and Maria Thun, *Gardening for Life.* See also Bernard Jarman, 'Harnessing the power of light', *Star and Furrow*, 107 (2007), 6–8.
6. John Wilkes, *Flowforms*, Floris Books, 2003.
7. See, for example, M. Guépin and R.T. Smith, 'Chickens: their nature and management', in *Star and Furrow*, 104 (2005), 4–8. Nutritional response to applied preparations can be quite rapid. For example, on a visit to New Zealand the author was shown data indicating increased butter fat of milk some 2–3 days following application of horn silica to pastures.
8. Planetary connections are discussed in depth by B. Lievegoed in a publication for the BDAA Experimental Circle, in 1951. Available from the BDAA.
9. Note 3 (1).
10. Manfred Klett, *Principles of the Biodynamic Spray and Compost Preparations*, Ch. 5, pp. 55–68, Floris Books, 2006.
11. Steiner, *Agriculture Course*, Lecture 5.
12. For general design, see note 3 (2).
13. I have been much moved by the lectures of Manfred Klett, and would therefore refer the reader again to note 10.
14. P. Tompkins and C. Bird, *Secrets of the Soil*, Viking Arkana, 1989.

15. Maria Thun, *Gardening for life*, Hawthorn Press, 1999.
16. Note 3 (2).
17. Alex Podolinsky, *Biodynamics: Agriculture of the Future*, Powelltown, Victoria, 2000. See also www.demeter.org.au/.
18. As such it has a resemblance to a temple offering or *puja*, as practised by Hindus and Buddhists.
19. *Hugo Erbe's New Biodynamic Preparations*, Mark Moodie Publications, Oaklands Park, Newnham-on-Severn, Glos.
20. Steiner, *Agriculture Course*, Lecture 4. This is produced by Greg Willis and Hugh Lovel in the USA, who have also developed homoeopathic preparations. Further versions of horn clay have been produced by Enzo Nastati at L'Albero della Vita, Italy. See *Commentary on Dr Rudolf Steiner's Agricultural Course*, p. 66.
21. Mark Moodie, *Star and Furrow*, 107 (2007) 9. Though not difficult to make, this preparation is available for example from www.biodynamics.net.au.
22. Steiner, *Agriculture Course,* Lecture 4.
23. This recipe appears in John Soper, *Biodynamic Gardening*.
24. See *Homoeopathy for Farm and Garden* by Vaikunthanath Das Kaviraj, Mark Moodie Publications, Oaklands Park, Newnham-on-Severn, Glos. (2006). See also *Introduction to the Homeodynamic method of agriculture, Homeodynamic Cultivation Handbook*, and *Foundations for a Development of Potentization*, by Enzo Nastati and his team. All are obtainable through homeodynamics@considera.org or eureka@spin.it. And see article by Michael Atherton in *Star and Furrow*, 109 (2008), 32–3.
25. Glen Atkinson has originated over 40 homoeopathic mixed preparations of which a small selection are available in the marketplace. See www.bdmax.co.nz. These are available in the UK through www.considera.org. A further range of commercial homoeopathic preparations is marketed in the USA by Greg Willis and Hugh Lovel. The inspiration for all these developments derives substantially from the work of Lili Kolisko who had worked for a time with Rudolf Steiner.
26. Those seeking further information or supplies should contact the respective websites. The various ranges of Eureka Institute (Enzo Nastati) products are obtainable by first becoming a subscription member which supports ongoing research activity.
27. These new 'homoeopathic or homeodynamic preparations' are more correctly described as 'potentized' or 'pre-potentized' because dilution in water is all that is necessary before application. Recently X500 and X501 have become available from the Josephine Porter Institute in the USA. See www.jpi.biodynamics.org.
28. A view held, for example, by Alex Podolinsky. See *Biodynamic Growing*, No. 2, 2004, Biodynamic Farming Association of Australia.
29. E.E. Pfeiffer, *Soil Fertility Renewal and Preservation*, Lanthorn Press.
30. Ingo Hagel's work can be accessed through the BDAA website and, in the original German, in the BDAA Library. Similar work has shown this effect with homoeopathic preparations of various kinds.
31. Lynne Carpenter Boggs et al., 'Effects of biodynamic preparations on compost development', *Biological Agriculture and Horticulture*, 17: 313–28 (2000). Also, 'Organic and Biodynamic management: effects on soil biology', *Soil Science*

Society of America Journal, 64: 1651–1659 (2000).

32. H.H. Koepf, *Research in Biodynamic Agriculture: Methods and Results*, Biodynamic Farming and Gardening Association, USA, 1993. See this publication for the work cited here by Speiss, Ahrens and von Wistinghausen.
33. Joachim Raup and Uli König, available from BDAA Library (English). See Uli König, 'On the biodynamic preparations', *Star and Furrow*, 98 (2003), 5–10. Also see Lynne Carpenter-Boggs et al., 'Biodynamic preparations: short-term effects on crops, soils and weed populations', *Biological Agriculture and Horticulture*, 15: 110–18 (2000).
34. Maria Thun also discusses evidence for the working of the preparations in her research publication *Results from the Biodynamic Sowing and Planting Calendar*, Floris Books, 2003, pp. 111–32.
35. J.P. Reganold et al., 'Soil quality and financial performance of biodynamic and conventional farms in New Zealand', *Science*, 260 (5106): 344–349.
36. L.L. Burkitt et al., 'Comparing irrigated biodynamic and conventionally managed dairy farms. 1. Soil and pasture properties', *Australian Journal of Experimental Agriculture* 47 (5): 479-488 (2007).
37. B. Pettersson and E. von Wistinghausen, *Organic, Biodynamic and Conventional Cropping Systems: A Long Term Comparison*, Woods End Agricultural Institute (1988). P. Mäder et al., 'Soil fertility and biodiversity in organic farming', *Science* 296 (5573), 1694-1697 (2002).
38. Ahrens and Bachinger (1992), *Lebendige Erde* (dissertation research based on plot trials at Darmstadt, Germany).
39. Freya Schikorr, 'A comparison of different methods of stirring the biodynamic field preparations', *Star and Furrow*, 82 (1994), 12–16.
40. Uli König and Uwe Geier, 'Trials on methods of stirring preparations', *Star and Furrow*, 109 (2008), 4–7. See also note 25, Chapter 8.
41. Based on the long-term DOK trials near Basle, Switzerland, the essential FiBL publication *Organic Farming Enhances Soil Fertility and Biodiversity* compares organic, biodynamic and conventional systems. A selection of material is also available through the BDAA and Considera websites in the UK.
42. Weed and pest problems are addressed widely throughout the organic movement and information is readily available on the internet.
43. Steiner, *Agriculture Course,* Lecture 6.
44. Maria Thun, *Gardening for Life*, pp. 91–4. Enzo Nastati offers radical alternative views on these methods. See Nastati, *Commentary* (Meeting 20).
45. See Steiner, *Agriculture Course,* Lecture 8, and also Enzo Nastati, *Commentary*, pp. 282f.

Chapter 6

1. In the UK the most commonly used are Maria and Matthias Thun's *The Biodynamic Sowing and Planting Calendar* and Nick Kollerstrom's *Gardening and Planting by the Moon*. The American *Stella Natura* is also a favourite, particularly for its inspiring articles, but corrections for Eastern Standard Time are necessary. Astro-calendars are also produced by Brian Keats in Australia and by Enzo Nastati in Italy.
2. Nick Kollerstrom's *Farmer's Moon: The Evidence for Increasing Crop Yields Through*

Use of Cosmic Rhythms (to be published in 2010) represents a lifetime's research into the use of calendars and is particularly important for its critical discussion of the different zodiacs.

3. Norman Davidson, *Sky Phenomena*, Floris Books (very readable as a basic introduction to astronomy). See also Joachim Schultz, *Movements and Rhythms of the Stars*, Floris Books.
4. The sun's rotation is referred to by Steiner in his lectures on bees. It is also mentioned by Thun in her book *Results from the Biodynamic Sowing and Planting Calendar*, Floris Books. This synodic rotation can be estimated by reference to sunspots.
5. Steiner spoke on many occasions about the moon. See, for example, *Outline of Esoteric Science* and *Agriculture Course,* Lecture 6.
6. A relationship between lunar synodic period and meteoroid particles in the atmosphere and, in turn, between these particles and planetary rainfall began to be established in the early 1960s. For a review of the early work, see Dycus and Martin, *J. Atmospheric Sciences* 26: 782–784 (1969).
7. E. and L. Kolisko, *Agriculture of Tomorrow*, Kolisko Archive Publications, 1978.
8. M. Thun, note 4.
9. Steiner, *Agriculture Course*, Lecture 6.
10. Note 2.
11. On the matter of zodiacs, readers will find helpful the brief commentary by Ian Bailey in *Star and Furrow*, 109 (2008), 40–1.
12. Precession is due to the fact that the earth gyrates (wobbles) on its axis. As a result, at the vernal (spring) equinox the sun moves anticlockwise against the background of stars by 1° every 72 years. The cycle of precession—known also as the Platonic Year—takes 25,920 years (72 × 360).
13. Influences of constellations on formative processes of human organs are discussed by Willi Sucher in *The Living Universe and the New Millennium*, Lecture 2 (10 Jan. 1956), Astrosophy Research Centre, Anastasi, 1998, pp. 67–92.
14. Steiner pointed out that when particular constellations are covered by the sun or moon, the human being is forced to compensate for lack of these spiritual forces. Reference to this effect occurs in Steiner's lectures to workmen. See *From Beetroot to Buddhism*, Rudolf Steiner Press, pp. 206–8.
15. Steiner, *Mystery of the Universe*, pp. 83–4.
16. Kollerstrom, note 2, Kolisko, note 7, and also Agnes Fife, *Moon and Plant*.
17. Steiner, *Agriculture Course*, Lecture 6.
18. Steiner, *Agriculture Course*, Lecture 6, and see *Harmony of the Creative Word*, p. 22.
19. M. Thun and H. Heinze, *Mondrhythem im siderichen Umlauf und Pflanzenwachstum. Forschungsring für biologisch-dynamische Wirtschaftsweise*, Darmstadt 1979, 125 pp.
20. For this and further research into lunar planting, the reader is referred to Kollerstrom, note 2. Also, on the internet, go to Google, select 'Images' and then type in 'Thun Trigon'.
21. Nick Kollerstrom has, however, much evidence of the importance of the lunar cycle for horse breeding. Note 2.
22. M. Thun, note 4.
23. M. Thun, note 4.

24. Steiner, *From Beetroot to Buddhism*, Rudolf Steiner Press, 1999, pp. 206–7.
25. Steiner, *Mystery of the Universe*, p. 58.
26. Kollerstrom, note 2.
27. Ian Bailey, note 11, and also see *Star and Furrow*, 110 (2009), p. 40.
28. See Enzo Nastati, *Commentary* (Lecture 2, Meeting 5).
29. Steiner, *Harmony of the Creative Word*, p. 22.
30. This was discussed in Chapter 2.
31. M. Thun, *Sowing and Planting Calendar* for 2008, p. 58. The six variants of Mirabelle plum have been witnessed by the author.
32. M. Thun, note 4, pp. 46f.
33. The paper was first impregnated with silver nitrate, dried and then followed by a caustic soda extract of the horn manure (see Chapter 10 for further reference to this method).
34. Reference to chromatographic and crystallization methods will be made in Chapters 8 and 10.
35. R. Hauschka, *Nutrition*, Ch. 21, pp. 142–3.

Chapter 7

1. A landmark book of its time was *Seeds of the Earth: A Private or Public Resource?* by P.R. Mooney, 1979.
2. See www.corporatewatch.org.uk.
3. For the UK, Defra sponsors the Centre for Organic Seed Information, the online database for which is managed by the Soil Association, Garden Organic and NIAB. For availability of organic seed, go to organicXseeds at www.cosi.org.uk. As a consequence of increased availability of organic seed, the market has reached a point where it is increasingly difficult to source untreated seed of conventional origin.
4. Requirements are detailed in the Defra compendium of standards and the Production Standards for Demeter certification, as periodically revised.
5. See Garden Organic's Heritage Seeds Project.
6. The Biodynamic Agricultural Association's Seed Development Project has been part of a Europe-wide movement to promote Demeter seeds. All suppliers of Demeter-quality seed, including Stormy Hall Seeds, Bingenheimer Saatgut, Vitalis, De Bolster Seeds, Sativa and Arcoiris, are listed at www.biodynamic.org.uk (click on 'seed project'). See also *Star and Furrow*, 109 (2008), 38–9, 'Producing biodynamic seeds in the UK', by Bernard Jarman.
7. It should be added that formation of viable seeds and fruits can occur in the absence of fertilization (apomictic species) while fruits can be formed without seed (parthenocarpy). Bananas, for example, exhibit vegetative parthenocarpy. It is therefore possible to uncouple fruit and seed formation from the fertilization process.
8. Pollen grains were referred to by Steiner as 'little vessels of warmth'.
9. Rudolf Hauschka, *Nutrition*.
10. The picture presented by Steiner of the cambium in relation to seed formation comes not in the agriculture lectures but from *Cosmic Workings in Earth and Man*, pp. 65–9, while the idea of a seed 'chaos' comes from Lecture 2 of the *Agriculture*

Course. This was clearly a subject about which Steiner felt deeply. He concludes the remarks in *Cosmic Workings* by stating: 'The earth first gives up her life to the plant, the plant dies, the air environment along with its light once more gives it life, and the cosmic spirit implants the new plant form. This is preserved in the seed and grows again in the same way . . . One sees in the growing plant how the plant world rises out of the earth, through death, to the living spirit.' Again, in his *Calendar of the Soul* for week 15, part of the entry reads: 'In all vegetation there is a brooding awareness of its own perpetual renewal. The seed does not come to the plant from outside but is created within by the forces bestowed by the wisdom of the cosmos.'

11. A very useful source on seeds is *The Seed Savers Handbook* by Jeremy Cherfas and Michel and Jude Fanton, Grover Books, 1996.
12. A comprehensive treatise is that of Raymond George, *Vegetable Seed Production*, CAB International, cabi@cabi.org.
13. Plant breeding techniques and their suitability for organic production were earlier reviewed by the Louis Bolk Institute. See their 1999 report *Sustainable organic plant breeding: vision, choices, consequences and steps*, www.louisbolk.nl. The biodynamic grower is advised to consult *Seeds of Concern* by Peter Brinch, on the Biodynamic Seed Development Project website, regarding non-permitted hybrids.
14. Carol Deppe, *Breed Your Own Vegetable Varieties*. Chapter 9 offers thoughts on saving seed from hybrids.
15. A helpful and accessibly written paper is 'Genome scrambling—myth or reality?' by Alison Wilson, Jonathan Latham and Ricarda Steinbrecher, *EcoNexus Technical Report*, October 2004, www.econexus.info.
16. The Soil Association publication *Seeds of Doubt* (2002) is an excellent review of North American farmers experiences of GM crops.
17. Martha L. Crouch, *How the Terminator Terminates: an explanation for the non-scientist of a remarkable patent for killing second generation seeds of crop plants*, Edmonds Institute, 1998.
18. Information on the consequences of GM technology can be easily accessed via the web. The incomparable publication *Seeds of Deception* by Jeffrey M. Smith (Green Books, 2004) is essential reading. I also found the treatment of GMOs by Adrian Myers in his *Organic Futures* comprehensive and stimulating.
19. The greater concern about GM products is the deception which lies behind their promotion and the obfuscation or denial which is patently evident whenever there is investigation of problems with these products. There are plenty of examples to choose from, a classic case being that of the L-tryptophan food supplement where many deaths and permanent disability occurred—see www.soilassociation.org.uk and click on 'library'.
20. Indigenous varieties of maize in Mexico—the centre of genetic diversity for maize—have for some years been contaminated with GM genes while in 2008 it was found that 50% of organic maize in northern Spain was genetically contaminated.
21. *The Seed Savers Handbook* (see note 11) offers in its 'Appendix A' a very useful 'pollination and seed storage table'. See also www.biodynamic.org.uk. 'Bio-

dynamic horticultural seed production' by Peter Brinch. This is a guide for Demeter out-growers producing for Stormy Hall. It gives pollination details, isolation distances and likely seed yields per sq. metre. With peppers, up to 68% cross-pollination has been observed in India.

22. Permitted fertility inputs to organic agriculture are 'approved' if they meet criteria concerning residue levels and environmental impact.
23. Recommended sources include *Breeding Field Crops* by D.A. Sleper and J.M. Poehlman, 5th edition, Blackwell, 2006, and *Vegetable Seed Production* by Raymond A.T. George. I am grateful to Tobias Mager for earlier instruction on this topic.
24. Dorian Schmidt, 'Observations in the field of formative forces in nature: methods and results' (transl. from German). Available from the Biodynamic Agricultural Association.
25. Plant diseases can be seed-borne (carried from generation to generation), soil-borne (surviving from one season to the next), carried by the wind or even introduced by contractors' vehicles and machinery.
26. A previous experiment has suggested that organic seed is more resistant to heating and remains viable.
27. Steiner, *Agriculture Course*, Lecture 2, p. 36, American edition. In essence, the 'cosmic' seed has to cease its dream and now unfold its capacities in the physical world.
28. In some soils we may be talking of water with sufficient dissolved oxygen rather than air as such.
29. These toxic chemicals, including copper and mercury compounds (the latter largely outlawed), have had harmful effects on a variety of wildlife, notably birds, and their legacy is to depress the activity of soil micro-organisms.
30. Seeds take up water more rapidly in the days leading up to full moon. The later part of the moon's ascending period might also be used to advantage (see Chapter 6).
31. Sattler and von Wistinghausen, *Biodynamic Farming Practice*, BDAA.
32. Chris Hull, 'Seeds and seed saving', Ch. 11 in *Biodynamics: New Directions for Farming and Gardening in New Zealand*, Random House, 1989.
33. *Homeodynamic Cultivation Handbook*, L'Albero della Vita, eureka@spin.it (20 seed-bath products available). See *Hugo Erbe's New Biodynamic Preparations*, pp. 74–6 (two seed-baths and one root dip), www.moodie.biz.

Chapter 8

1. Inevitably there are questions about the meaning of 'vital energies'. 'Radiant forces' merely expresses another aspect of what is meant here. Such energies have been referred to throughout this book.
2. Viktor Schauberger's work is well reviewed in *Living Energies* by Callum Coates, Gateway Books, 1996. Comments by Enzo Nastati are very helpful on this subject. See his *Commentary on Dr Rudolf Steiner's Agriculture Course*, pp. 108–9, 138.
3. Countless books have been published on hydrology. The one I earlier made most use of was *Hydrology*, ed. Oscar E. Meinzer, Dover Books, 1949.
4. Those wishing to learn more about the chemistry of water should visit Martin

Chaplin's excellent website www.lsbu.ac.uk/water/.

5. Despite the importance of this chemical property of water, the actual amount ionized at any moment is minute while each ion has a probability of existence of only about 10^{-12} second.
6. The scale of pH is logarithmic, extending from 0 to 14. Neutrality has the value 7. Below 7 is acidic, above 7 alkaline. These variations arise from water's reaction with minerals, gases and organic substances.
7. Theodor Schwenk, *Sensitive Chaos*, Rudolf Steiner Press. E. and L. Kolisko, *Agriculture of Tomorrow*, Kolisko Archive Publications, 1978.
8. Samuel C.F. Hahnemann, *Organon der rationellen Heilkunde* ('The Organon of Rational Medicine'), 1810.
9. According to Steiner, memory exists within the human *etheric* body. As water is clearly connected to life processes, this predicates water as having etheric quality.
10. Theodor Schwenk, *The Basis of Potentization Research.*
11. Enzo Nastati, *Introduction to the Homeodynamic Method of Agriculture*, L'Albero della Vita.
12. R. Hauschka, *The Nature of Substance*, 1966, Ch. 17.
13. The word 'resonance' or even 'etheric resonance' is preferred to 'imprint' as such patterning is not permanent.
14. N. Grant, M. Moodie, and C. Weedon, *Sewage Solutions*, Centre for Alternative Technology, Machynlleth, 1996. See also www.considera.org.uk.
15. For example, the village of Nun Monkton in Yorkshire, UK.
16. Masaru Emoto, *The Message from Water*, 2004.
17. Theodor and Wolfram Schwenk, *Water—The Element of Life*, Anthroposophic Press.
18. Alan Hall, *Water, Electricity and Health*, Hawthorn Press, 1997.
19. I refer here to *Occult Chemistry* by the theosophists Charles W. Leadbeater and Annie Besant, 1951. Reviewed by the physicist Steven M. Phillips in his *Extrasensory Perception of Quarks.* It is also abundantly evident that Lucretius, in *De rerum natura* (first century BC), was able to mentally visualize the nature of atoms.
20. P. Tompkins and C. Bird, *Secrets of the Soil*, 'Vortex of life', Viking, 1991, pp. 99–115.
21. Drawn mainly from Rudolf Hauschka's *Notes for the BDAA Experimental Circle*, 1949.
22. I am grateful to the late Pauline Anderson for having made this observation in a talk to the Experimental Circle of the BDAA.
23. Enzo Nastati, *Introduction to the Homeodynamic Method of Agriculture*, p. 19.
24. For information on flowforms and their applications, see *Flowforms: The Rhythmic Power of Water*, by A. John Wilkes, Floris Books, 2003. Contact the Healing Water Institute, Emerson College, Forest Row, E. Sussex, RH18 5JX. For flowform design and installation, contact Ebb and Flow Ltd., Ruskin Mill, Nailsworth, Gloucestershire, GL6 6DQ.
25. See *Flowform Water Research 1970–2007*, publication of the Healing Water Institute (2008), www.healingwaterinstitute.org.
26. E. Davenas, J. Benveniste, et al., 'Human basophil degranulation triggered by

very dilute antiserum against IgE', *Nature*, 333, 1988, p. 816.

Chapter 9

1. Steiner, *Agriculture Course*.
2. Karl König, *Earth and Man*, Biodynamic Literature, 1982. See also reference to Karl König in Chapter 11.
3. This connects with the question of traditional reverence mentioned in Chapter 1.
4. Spinosa's ways of seeing and Goethe's journey, Chapter 4 of Rudolf Steiner's *Goethean Science*, Anthroposophic Press.
5. The nature of the Christ Being, as distinct from the person of Jesus of Nazareth, is explained in many of the lectures and books of Rudolf Steiner.
6. Further information about Pishwanton is accessed through the website www.pishwanton.com or by writing to Margaret Colquhoun at: The Life Science Trust, Old Bank Building, 1B, High Street, Gifford, East Lothian, EH41 4QU.
7. Christopher Day, *Consensus Design*, Elsevier, 2002.
8. Articles relating to Pishwanton by Margaret Colquhoun include 'Healing the land, healing ourselves', *Caduceus*, 72 (2007), 6–11, and 'Pishwanton's herb garden', *Star and Furrow*, 110 (2009), 36–9. A future publication will cover the garden and woodland design process. Donations towards the completion of the Goethean Science Building will be gratefully received (see note 6).

Chapter 10

1. For example, J. Humphreys, *The Great Food Gamble*, Hodder, 2001, and E. Schlosser, *Fast Food Nation*, Penguin, 2001.
2. An excellent overview of this subject is provided in the Soil Association's publication *Organic Farming, Food Quality and Human Health*, 1993.
3. I.G. Plaskett, 'Nutritional therapy to the aid of cancer patients', *Int. J. Alternative and Complementary Medicine* (Dec. 1999).
4. Wendy E. Cook, *The Biodynamic Food and Cookbook*, Clairview Books, 2006.
5. Major supermarkets employ tasting panels. This particular observation from 1998 derives from Marks and Spencer's.
6. Virginia Worthington, 'A fresh look at an old debate—is organic food more nutritious?' *Acres* (USA), June 1998, p. 17. Antioxidants study, 'Quality low input food', at Newcastle led by Professor Carlo Liefert as reported in *Green Health Watch*, 34 (2008), p. 10. For changes in nutrient content, see Anne-Marie Mayer, 'Historical changes in the mineral content of fruits and vegetables', *British Food Journal*, 99 (6), 1997, 207–211. See also *Organic eprints* on the FiBL website.
7. Australian exports reported by Alex Podolinsky. Ghana information gathered by the author while on a DfID project in 1998. For Pettersson's research and other work on keeping quality including that of Samaras quoted below, see H.H. Koepff, *Research in Biodynamic Agriculture*, 1993.
8. *Organic Farming, Food Quality and Human Health* (note 2), pp. 49–51. See also Paul Mäder et al., 'Wheat quality in organic and conventional farming', *J. Sci. Fd. Agric.*, 87 (2007): 1826–1835.
9. R.D. Hodges and A.M. Scofield, 'Effect of agricultural practices on the health of

plants and animals produced—a review', in *Environmentally Sound Agriculture*, ed. W. Lockeretz, Praeger, 1983. Also note 2.

10. For example, J. Auger et al., 'Decline in semen quality among fertile men in Paris in the last 20 years', *New England J. Medicine* (1995). See also 'High sperm density among members of organic farmers organisation', *Lancet*, 343 (1994), 1498, and S. Wijeratna et al., 'Semen quality of men in organic and conventional tea plantations, and in an urban population of Sri Lanka', *Proc. Sri Lanka Inst. of Biol.*, 2003.
11. Steiner refers to such an effect when discussing animal husbandry in Lecture 8 of *The Agriculture Course*. In another context, see Buchmann and Hiss, 'A comparison between hydroponics and biodynamic cultivation', *Star and Furrow*, 97 (2002), 15–17.
12. It is generally known that mothers who are poorly nourished give birth to smaller children, but this research shows that circumstances experienced by either sex of a particular generation confer a predisposition on grandchildren rather than to immediate progeny—an epigenetic effect. This has resonance with later discussion of health problems. See Gunnar Kaati, 'Epigenetic inheritance in Man', www.integratedhealthcare.eu/images/stories/news/Kaati.pdf.
13. Article by Dr Allen Buresz and the references which follow, www.all-natural.com/add.html. The latter examples were drawn from the *Newsheet* of the Experimental Circle of Anthroposophical Farmers and Gardeners, Sept./Oct. 1997. See IQ and diet by D. Benton, *Nature*, 255 (6362) (1992), p. 667, and 'Nutrition—the brainpower advantage', www.healingwithnutrition.com/adisease/add-adhd/harrelstudy/html. For school meals see www.telegraph.co.uk/education/educationnews/4423132/jamieoliver.
14. K. Huber et al., 'The Monastery Study—How Does Food Quality Affect Body, Soul and Spirit?', Darmstadt 2003. Translation from German available on BDAA website.
15. These were fairly and objectively assessed in a Soil Association publication (see note 2). See www.biodynamic-research.net.
16. E. and L. Kolisko, *Agriculture of Tomorrow*, Kolisko Archive Publications.
17. E. Pfeiffer, *Chromatography Applied to Quality Testing*, Bio-Dynamic Literature.
18. For example, E. Pfeiffer, *Sensitive Crystallization Processes*, Anthroposophic Press. This technique requires very exacting laboratory controls to achieve reliable results. For example, in the slow course of crystal formation, the slightest vibration will tend to cause crystallization in super-saturated solutions.
19. U. Balzer-Graf, 'Vital quality—quality research with picture forming methods', *Star and Furrow*, 94 (2000), 17–21.
20. See Chapter 6 for an example of an eclipse. Agnes Fyffe in particular has conducted chromatograms over periods of cosmic change. See *Moon and Plant*, Society for Cancer Research, Arlesheim, 1975.
21. J. Kahl et al., 'First tests on standardized biocrystallization on milk and milk products', *Eur. Food Res. Technol.* (2009) 229: 175–178.
22. For an enlargement on this theme see Wendy Cook, 'Food vitality', *Star and Furrow*, 101 (2004), 31–35. R. Hauschka, *Nutrition*. This work and the two

volumes by Gerhard Schmidt, *The Dynamics of Nutrition* and *The Essentials of Nutrition* (both published by Bio-Dynamic Literature), are helpful for those wishing to follow this subject in greater depth.

23. Alan Hall, *Water, Electricity and Health.*
24. R. Hauschka, *Nutrition*, p. 123.
25. The nutritional value of rice is seriously affected by milling. Along with the introduction of modern varieties, the trend to white rice in poorer countries initially led to malnutrition and beri-beri (particularly in children), due to loss of vitamin A.
26. Dr Robert McCarrison was a nutritionist and contemporary of Albert Howard and Eve Balfour.
27. Gustav von Bunge, 'A handbook of physiological and pathological chemistry', 1887 (in German). See also Steiner, *Bees*, Lecture 3, p. 42.
28. Mae-Wan Ho, *The Rainbow and the Worm—The Physics of Organisms*, Institute of Science in Society. See the article by Strube and Stolz in *Star and Furrow*, 105 (2006), 30–2. For German readers Marco Bischoff's book *Biophotonen. Das Licht in Unseren Zellen, Gebundene Ausgabe* ('Biophotons—the light in our cells'), Gebundene Ausgabe, should be consulted. See internet.
29. Many have digestive and other experiences—for example poor sleep—after eating certain types of meat. This is connected with the animal having had conscious experiences in its life. While a step further, those who have received organ transplants sometimes report mental experiences not apparently belonging to them, as if a residue of the donor remains.
30. A brief statement of Steiner's position is found in the *Agriculture Course,* Ch. 4, p. 63. A wide selection of Rudolf Steiner's lectures on nutrition are included in *Nutrition and Stimulants*, compiled by K. Castelliz and B. Saunders-Davies (1991). See also Steiner, *Nutrition—Food, Health and Spiritual Development*, Rudolf Steiner Press, 2008. Wendy Cook, in *Foodwise*, offers a wide-ranging and highly accessible picture of nutrition which can be recommended as a good starting point. In addition to the volumes by Hauschka and Schmidt, the reader should also consult *World Ether*, extracts from Rudolf Steiner compiled by E. Hagemann, pp. 86–7.
31. The cellular metabolism underlying this operates according to the tricarboxylic acid or Krebs cycle.
32. Mineral salts were mentioned in the context of the nutritional value of animal and human manures in Chapter 3. This topic will be discussed later in the present chapter.
33. In Chapter 4 we discussed light and those formative forces which it contains. Arising from light, as it works in the plant, are sugars and carbohydrates, which provide energy for the physical body and formative forces for the etheric body.
34. This dematerialized blueprint is a structure borne by carbon, the form-bearer.
35. Hagemann has calculated that 20–35g of cosmically derived nourishment is required daily in order to replace bodily substances which are lost (see his compilation of extracts from Rudolf Steiner, *World Ether*, p. 48). A figure of around 25 grams assimilated from every 1 kg of food ingested is supported by nutritionists Petra Kuhne and Maurice Orange. *Nutrition and Stimulants* (see note

30) is helpful (especially Chapters 17 and 18).

36. Vitamins appear to play a role in this, for we know vitamin D connects with sunlight. In certain cases, notably adepts, people appear to have lived off light alone. See for example, M. Werner and T. Stöckli, *Life from Light*, Clairview Books, 2007. In Chapter 3 it will be recalled that we mentioned different pathways of planetary forces received by the animal organism.
37. R. Hauschka, *Nutrition*, Ch. 4.
38. Karl König, *Earth and Man*, Bio-Dynamic Literature, 1982.
39. Here, we might draw a parallel with the different results obtained from the chromatographic and crystallization picture-forming methods!
40. This knowledge allows fresh perspective on certain physiological problems. For example, on long expeditions, in addition to continuous exertion and limited supplies, a contributory factor to muscle wastage may well be the intake of preserved food with low 'vital force' content.
41. *Nutrition and Stimulants* (see note 30, Ch. 17).
42. Steiner, *Agriculture Course*, Lecture 8 for an introduction to this concept.
43. Wendy Cook in *Foodwise* (Chapters 14 and 15) discusses the value of minerals in nutrition.
44. Higher proportions are evident in some other countries, the figure currently being 1 in 3 couples in Sri Lanka.
45. Steiner made the point that reproductive energy was an enhanced level of metabolism. As every individual's system is unique it is clear that these problems will manifest especially where a genetic factor applies and where there has been an unhelpful family history of diet and health.
46. The causes of low sperm counts or other aspects of male infertility are complex, but stress and environmental chemicals play a large part. See www.babyhopes.com/articles/causes-low-sperm-count/html and many other internet sources.
47. Early research showed a relationship between levels of lindane in fatty tissue and women with breast cancer. Lest these remarks be considered to have only a narrow frame of reference, it has been estimated that at least 70,000 suffer the effects of agrochemicals in India while around 10,000 deaths occur annually in the Third World due to chemicals, many of which have been banned elsewhere.
48. The 'generational effect' is not without clinical precedent. The thalidomide tragedy is well known, while use of DES by women in the 1950s and 1960s had consequences for their female offspring.
49. There is a particular approach to cancer treatment adopted by anthroposophical therapists. For details of this, the booklet by Udo Renzendrink, *Diet and Cancer* (Raphael Medical Centre/Rudolf Steiner Press, 1988) is suggested as a first source.
50. 'Iodine: high-dose intake cuts breast cancer and hypothyroid-related illnesses', *Caduceus*, 75, 18–21. This article, by Donald W. Miller, a cardiac surgeon and professor of surgery at the University of Washington, Seattle, is of great potential importance.
51. This could be as blinkered as a 'glucose-only' view of the role of insulin. See for example www.vivo.colostate.edu/hbooks/pathphys/endocrine/pancreas/

insulin_phys.html.

52. We are encouraged to reduce total salt intake for very good reasons which need not be repeated here. Some salt contains anti-caking agents while others reduce the sodium levels, but there is no current guidance on minimum levels of iodine as iodide. Iodine is, however, available in tablet form, see Renzendrink, *Diet and Cancer* (note 49).
53. As a result of such dietary stresses, there may be a tendency for proto-cells to preferentially form fat cells.
54. We refer here to so-called 'diet' versions of soft drinks. See 'Aspartame and weight loss', www.rense.com/general3/asper.htm. The internet provides many references.
55. Such a factor might be epigenetic—involving changes in gene expression rather than of DNA sequences.
56. 'Diabetes rates are increasing among youth', www.nih.gov/news/pr/nov2007/niddk-13.htm.
57. Brian Goodwin, *Nature's Due*, p. 169.
58. P. Lachmann, et al., *Clinical Aspects of Immunology*, Wiley, 5th ed., 1993.
59. The thymus gland, which lies just behind the breastbone, plays an important part in the body's defences. T-lymphocytes (white cells) produced in the bone marrow migrate to the thymus where they mature and are stored. This gland *should* be relatively large in young people while it shrinks in size as one gets older.
60. R.K. Chandra, 'Nutrition and the immune system: an introduction', *Am. J. Clin. Nutr.*, 66 (2): 460–462 (1997). E.S. Wintergerst, et al., 'Contribution of selected vitamins and trace elements to immune function', *Ann. Nutr. Metab.*, 51 (4), 301–323 (2007).
61. Around 40% of British people now suffer from allergies while the incidence of children diagnosed with food allergies has tripled in the past decade. Diet is being recognized as part of this picture, *Daily Telegraph*, 15 April 2009. It is suggested the reader consults the article by Nick Jones in *Star and Furrow*, 110 (2009), 23–6, 'A new approach to milling and baking taking account of biodynamic principles'.
62. We focus on chickens here as avian influenza now presents a real threat to the human population. Equally serious is the factory farming of pigs, especially in the light of recent outbreaks of swine flu.
63. Methicillin-resistant *Staphylococcus aureus*. See www.defra.gov.uk/animalh/diseases/zoonoses/mrsa, and insert 'animal-related' or 'community-acquired MRSA' into an internet search engine.
64. R. Audry (1996), 'Homeopathy and chronic fatigue', *Int. J. Alt. Medicine*, 14: 12–16. This research reports an impressive outcome as a result of a double-blind trial.
65. Opposition to vaccination is reviewed in a booklet by Alan Phillips entitled *Dispelling Vaccination Myths*', published by Prometheus, 55 Hob Moor Drive, Holgate, York YO24 4JU.
66. Hauschka states that if there are unresolved obstacles to be dealt with by the young person, the individuality may not be able to incarnate properly, leading to a disturbed soul-world, just as a crooked mirror gives a distorted reflection (*Nutrition*, p. 49). This adds potency to the statement of Juvenal (*c.* AD 60–*c.* 130):

'Orandum est ut sit *mens sana in corpore sano*' (One should pray for *a sound mind in a sound body*), *Satires* no. 10, l, 356.

67. Organophosphates were used in this case to kill insects that could have transmitted disease among those living in close proximity. For consequences of their use in British agriculture, see *Animal Pharm* by Mark Purdey, Clairview Books, 2007.
68. Nick Anderson, *Green Health Watch*, No. 34, p. 12, referring to the research of Dr Sarah Myhill.
69. It is increasingly unwise to plan meals without enquiring about special dietary needs, let alone dietary preference. Such problems do not make it easier to establish sociable family eating patterns.

Chapter 11

1. Trauger Groh and Steven McFadden, *Farms of Tomorrow* and *Farms of Tomorrow Revisited*, Biodynamic Farming and Gardening Association Inc., USA.
2. Rudolf Steiner, *Agriculture Course*.
3. See Chapter 2, Note 10.
4. As the quotation originally referred to England, readers will perhaps allow the changing reference to Britain and England!
5. For further information concerning the Camphill Village Trust, visit www.camphill.org.uk. Camphill was named after the Camphill Estate in Aberdeen where the first home for children with special needs was established in 1940.
6. Rudolf Steiner was actively engaged in promoting what he termed 'the threefold commonwealth' during the years 1917–22. He gave many lectures on the theme and sought to inform many of the leading figures of his time. It met with both interest and opposition. In the end it could not be realized and instead the notorious Treaty of Versailles was foisted on Europe. This led ultimately to the rise of National Socialism, the Second World War and the subsequent partition of Europe.
7. According to figures held by Co-operatives UK, 50% of the world's population is today involved in one way or another with co-operatives.
8. In the UK, establishments involving care in the community are supported on a per capita basis by government funding to local authorities. The word 'intentional' signifies being created as an entity.
9. For further information visit www.cuco.org.uk. See the Soil Association's publication *A Share in the Harvest*, ISBN 0 905 200 82 9.
10. For further information visit www.stroudcommunityagriculture.org. SCA operates at two sites as is explained later.
11. For further information visit www.tablehurstandplawhatch.co.uk. Tablehurst and Plaw Hatch Community Farm is based in and around the neighbouring villages of Forest Row and Sharpthorne in East Sussex. It is set up as an Industrial Provident Society.
12. For information visit www.fordhallfarm.com.
13. This is part of The Cultural Freedom Trust. See www.biodynamics.org.uk and www.bdlandfund.org.
14. Rob Hopkins, *The Transition Handbook—From Oil Dependency to Local Resilience*,

Green Books, 2008. This was reviewed in *Caduceus*, 75 (2008), 32–33. For information on the diversity of activities involved, see www.transitionculture.org. At the time of writing, more than 600 communities are at various stages of progress along this path while the movement is spreading to other countries.

Chapter 12

1. In the UK, prolonged under-investment means that public transport alternatives are beyond attainment within an affordable framework, thus making the country seriously disadvantaged when a real energy crisis occurs.
2. The effects of global warming have been more noticeable since the 1990s. British weather now displays elements of tropical ferocity, with intense rainfall events and tornados becoming a regular occurrence. The tragedy of droughts in Australia's Murray-Darling basin, Africa and elsewhere, and of increasingly serious forest fires, is a reminder of the fragility of many of the world's land resources.
3. Indiscriminate use of antibiotics in intensive animal farming systems is one reason why antibiotic-resistant bacteria will pose a worsening problem (see Chapter 10).
4. I refer here to the creative forces that have brought the world into existence and sustained us to this point (Chapter 2).
5. Peter Taylor, 'Inconvenient truths expose flawed CO_2 model', *Caduceus*, 76, 2008, 6–12. A highly documented article explaining the current uncertainties. See www.ethos-uk.com.
6. For the advantages of organic farming, see Chapter 1. Connections with cosmic forces are explored in Chapter 4.
7. *Organic Farming and Climate Change*, International Trade Centre, 2008. Adrian Myers, *Organic Futures*, 2005, also gives an excellent overview of these issues.
8. In this context, the role of the worldwide Permaculture Association is regarded as of great importance. The benefits of mixed cropping were discussed in Chapter 1.
9. See below, and Chapter 11.
10. In this respect 'sell-by' dates bear as much relation to issues of stock turnover and consumer consumption as they do to food safety.
11. While food is imported from wherever *production costs* are lowest, this can apply simply to the processing stage. For example, prawns landed in western Scotland are sent to China for peeling! Because they are harvested and packed in the UK, that is what appears on the packaging.
12. Indeed, the policy turned a blind eye to our historical need of effective agricultural capacity, notably during the Napoleonic Wars and subsequently two World Wars.
13. It is reported that in Californian communities local nurseries are now stocking more vegetable seedlings than ornamental plants due to the explosion in backyard gardening.
14. In Spain, where 25% of maize is grown with GM seed, organically grown varieties were so contaminated in 2007 that they had to be destroyed, resulting in shortages of the organic option in supermarkets.
15. Lockeretz et al. (1984), 'Comparison of organic and conventional farming in the corn belt', *American Society of Agronomy Special Publication*, 46: 37–49. See also Myers, *Organic Futures*, Green Books, pp. 194–6. Thought-provoking data is

provided by Catherine Badgley et al. (2007), 'Organic agriculture and the global food supply', *Renewable Agriculture and Food Systems*, 22: 86–108.

16. A useful review of the achievements of organic farming overseas is to be found in *Organic Futures* by Adrian Myers, while the research of Jules Pretty is notable in this area.
17. There are over 31 million hectares registered as organic across the world (the UK has 700,000 ha). See: *The World of Organic Agriculture—Statistics, and Emerging Trends*, 2008, published by IFOAM and FiBL.
18. For the basis of this remark, see Chapter 11. Rights are normally secured by the state where, in a democratic system, those in power are elected by the people.
19. Among many sites, see for example www.natural-health-information-centre.com/codex-alimentarius and the book *Evil Empire* by Paul Hellyer.
20. The purchase of inefficient government-operated farms in Sri Lanka by CIC is an example of the current diversification of chemical companies into direct involvement with agriculture. Readers will appreciate that this is rather different from car manufacturers beginning to produce electric cars.
21. Lord Acton (1834–1902): 'Power tends to corrupt and absolute power corrupts absolutely.'
22. This will merely encourage the use of GM crops, engineered to be ready for subsequent digestion to produce ethanol.
23. On the internet one can find vitriolic criticism of the use of land in this way as a conspiracy against poor countries. For more rational treatment, see articles published by the Society for International Development at www.sidint.org/, for example Peter Rosset, 'Food sovereignty and the contemporary food crisis', *Development*, 51: 460–463 (2008).
24. The factor involved is claimed to be between 5 and 7 times. Ultimately we have to accept that the argument for animal grazing is problematic, as lucrative markets for meat products could lead to scarcity of land for local food crops, and in turn to food shortages, malnutrition and worse.
25. Martin Large and Neil Ravenscroft, article in the *Ecologist,* March 2009.
26. We should also acknowledge that in Britain and a number of other countries *all* farming is regulated, whether effectively or not, by environmental agencies.
27. In this context, consider the statement by Dr Christine Jones: 'Soil is the largest carbon sink over which we have control.' Also the US Department of Energy considers that 'enhancing natural processes offers the most cost-effective means of carbon sequestration'. A 1% increase of soil organic carbon in the top 33.5 cm of soil locks up approximately 100 tons of atmospheric CO_2 per sq. km.
28. While the impracticality of compost application over large areas is immediately obvious, the cost of establishment of green manure crops must also be reckoned with.
29. Walter Bagehot (1872) in *Physics and Politics*, 'The age of discussion'.
30. Despite the fact that no biodynamic or organic herds suffered from BSE, it has become impossible for those in BSE-affected areas to use their own animals' organs for the biodynamic preparations. Because of the perceived risks from nerve tissue, these materials have remained on a 'restricted' list. This matter was referred to in Chapter 3.
31. For those preparations requiring an animal organ, Maria Thun suggests to enclose

these in the bark of a tree having the same planetary connection as the original animal organ (*The Biodynamic Sowing and Planting Calendar,* 2009). See also Chapter 3, note 22 and the article by Bernard Jarman in *Biodynamic Newsheet*, March 2009. Due to the 'astral' quality engendered by the animal parts, vegetarian options cannot achieve the same effects as Steiner's originals; for Demeter certification, they are not permitted as substitutes.

32. The purchase of preparations may be undertaken without much thought in developed countries, but in some others their relative cost and the unreliability of postal services constitutes a significant deterrent.
33. Having an awareness of the overseas situation and of cultural issues provides a mirror in which to view biodynamics. It calls to mind the aphorism of Rudyard Kipling (*The English Flag*, 1892), 'And what should they know of England, who only England know?' Besides the compost preparations, concepts such as the earth's *seasonal* breathing process thus apply mainly to high latitude locations.
34. The stirring of preparations has been discussed in Chapters 5 and 8. Here, we should merely note that existing Demeter standards permit the choice between barrel, machine and flowform stirring despite the differences of opinion on their use.
35. The work of Atkinson and Nastati was referred to in Chapter 5. Nastati has reported a trial he and his team undertook in 1998 comparing his potentized compost preparation with the original compost preparations. See *Agriculture Dossier*, Eureka, pp. 36–7. I am not aware of any current study (2009) to compare the efficacy of alternative versions of the spray preparations.
36. In this respect, although it is important to maintain sets of standards, the *person* may be at least as important as the *practice*.
37. The views expressed in this section are the author's own, based on experience in the UK and overseas. His remarks should not be taken to imply that these matters are under wider discussion.
38. This was discussed in Chapter 11. Camphill Villages are 'internal' communities catering for those with special needs. Here, 'community support' is on two levels, the village unit itself and the local authority, Department for Social Security. By contrast, Garvald homes and day centres (around Edinburgh), although similarly inspired, offer a looser 'employee relationship' for their workers.
39. Many developments are now taking place—see internet under 'local agriculture'.
40. Helena Norberg-Hodge argues the need to support local communities, particularly those of developing countries, against the forces of globalization. See Further Reading. Also Jules Pretty, *Agri-Culture*.
41. www.forestschools.com.
42. Initiated by Aonghus Gordon, there are currently centres at Nailsworth (Ruskin Mill), Stourbridge (Glasshouse College), Sheffield (Freeman College) and Darlington (Clervaux Trust). The success of these schemes for 'educating back into the community' seems likely to lead to further initiatives in the course of time. See www.rmet.org.uk.
43. See note 14, Chapter 11.

Additional source information for the illustrations:

Fig 1.6 from drawings provided by R. Ulluwishewa. Fig 1.11 from J. Bockemühl, *In Partnership with Nature*, 1981. Fig 2.1 from W. Dewanarayana, *Cosmic Influences on Healthy Crop Production*, University of Peradeniya, Sri Lanka. Fig 2.3: unpublished diagram, John Wilkes, Emerson College, Sussex. Fig 2.4 courtesy www.dnr.sc.gov (with modification). Fig 2.5: R. McCracken, *Stella Natura Calendar*, 2001. Fig 2.6: Fraunhofer archive, as accessed for example on www.honolulu.hawaii.edu. Fig 2.7 courtesy www.spectralecology.org (with additions). Fig 2.11: redrawn from G. Grohman, *The Plant*, Vol 1. Fig 2.12 from B. Keats, *Betwixt Heaven and Earth*. Fig 2.14: see ch. 2, note 58. Fig 4.4 courtesy www.food-info.net. Fig 4.6 from J. Bockemühl, *In Partnership with Nature*, 1981. Fig 8.4 From T. and W. Schwenk, *Water, The Element of Life*, 1989. Fig 8.6 from A. Hall, *Water, Electricity and Health*, 1997. Figures in Chapter 9 are from the collection of M. Colquhoun. All other figures are the author's originals or photographs from his collection.

Further Reading

Works by Rudolf Steiner

Agriculture Course. Two editions are currently available. The original English translation by George Adams has been republished (2004) as *Agriculture Course: The Birth of the Biodynamic Method*, by Rudolf Steiner Press, UK. A more recent translation (1993) was made by Catherine Creeger and Malcolm Gardner, and published as *Spiritual Foundations for the Renewal of Agriculture*, by The Bio-Dynamic Farming and Gardening Association Inc., USA.

Agriculture: An Introductory Reader. Extracts from Rudolf Steiner compiled by Richard Thornton Smith with commentary and notes, Sophia Books, 2003

Bees, Anthroposophic Press, 1998

The Cycle of the Year as Breathing-process of the Earth, Anthroposophic Press, 1984

Harmony of the Creative Word. The Human Being and the Elemental, Animal, Plant and Mineral Kingdoms, Rudolf Steiner Press, 2001

Nutrition and Stimulants. Extracts compiled by Katherine Castelliz and Barbara Saunders-Davies, Biodynamic Farming and Gardening Association Inc., 1991

An Outline of Esoteric Science, Anthroposophic Press, 1997

World Ether—Elemental Beings—Kingdoms of Nature, extracts from Rudolf Steiner compiled by Ernst Hagemann, Mercury Press, NY 1993

Other authors

Cook, Wendy, *Foodwise: Understanding What We Eat and How it Affects Us*, Clairview Books, 2003. See also *The Biodynamic Food and Cookbook*, Clairview Books, 2006

Davidson, Norman, *Sky Phenomena: A Guide to Naked-eye Observation of the Stars*, Floris Books, 1993

Edwards, Lawrence, *The Vortex of Life: Nature's Patterns in Space and Time*, Floris Books, second edition, 2006

Groh, Trauger and Steven McFadden, *Farms of Tomorrow Revisited: Community Supported Farms—Farm Supported Communities*, Biodynamic Farming and Gardening Association, Kimberton 2000

Hauschka, Rudolf, *The Nature of Substance*, and also *Nutrition*, both Rudolf Steiner Press

Henderson, Gita (ed.), *Biodynamic perspectives: Farming and Gardening*, New Zealand Biodynamic Association, 2001

Klett, Manfred, *Principles of the Biodynamic Spray and Compost Preparations*, Floris Books, 2006

Koepf, Herbert H., *The Biodynamic Farm*, Anthroposophic Press, 1989

Kolisko, Eugen and Lili, *Agriculture of Tomorrow*, Kolisko Archive Publications, 1978

König, Karl, *Earth and Man*, Biodynamic Literature, 1982

Myers, Adrian, *Organic Futures: The Case for Organic Farming*, Green Books, 2005

Nastati, Enzo, *Commentary on Dr Rudolf Steiner's Agriculture Course*, Eureka, Trieste 2009

Norberg-Hodge, Helena, *Bringing the Food Economy Home*, Zed Books, 2002

Podolinsky, Alex, *Biodynamic Agriculture Introductory Lectures* (2 volumes), Gavemer Publishing, Sydney 1985, 1989

Pretty, Jules, *Agri-Culture—Reconnecting People, Land and Nature*, Earthscan, 2002

Proctor, Peter with Gillian Cole, *Grasp the Nettle: Making Biodynamic Farming and Gardening Work*, Random House, Auckland 1997

Sattler, Friedrich and Eckard von Wistinghausen, *Biodynamic Farming Practice*, Biodynamic Agricultural Association, 1992

Schultz, Joachim, *Movement and Rhythms of the Stars: A Guide to Naked Eye Observation of Sun, Moon and Planets*, Floris Books, 1986

Soper, John, *Bio-dynamic Gardening*, Souvenir Press, 1996

Storl, Wolf D., *Culture and Horticulture*, Biodynamic Literature, 1979

Thun, Maria, *Gardening for Life*, Hawthorn Press, 1999

Wilkes, John, *Flowforms: The Rhythmic Power of Water*, Floris Books, 2003

Wilkinson, Roy, *Rudolf Steiner: An Introduction to his Spiritual World-view, Anthroposophy*, Temple Lodge, 2001

Wright, Hilary, *Biodynamic Gardening—For Health and Taste*, Floris Books, 2009

INDEX